Concepts in Architectural Acoustics

Concepts in Architectural Acoustics

M. DAVID EGAN

Consultant in Acoustics and Associate Professor of Architecture, Clemson University

McGRAW-HILL BOOK COMPANY

New York St. Louis San Francisco Düsseldorf Johannesburg Kuala Lumpur London Mexico Montreal New Delhi Panama Rio de Janeiro Singapore Sydney Toronto

Library of Congress Cataloging in Publication Data

Egan, M. David
Concepts in architectural acoustics.

Bibliography: p.
1. Architectural acoustics. I. Title.
NA2800.E34 1972 729'.29 72-000039
ISBN 0-07-019053-4

1234567890 HDBP 765432

The editors for this book were William G. Salo, Jr., and Robert E. Curtis, the designer was Naomi Auerbach, and its production was supervised by George E. Oechsner. It was set in IBM Selectric Baskerville and Univers Bold by Textart Service, Inc.

It was printed by Halliday Lithograph Corporation and bound by The Book Press.

Contents

SECTION 2 Sound Absorption 23

SECTION 3 Sound Isolation 53

Foreword

The need for training in the environmental sciences for those who aspire to work in the building professions has never been more urgent. This need will continue for the foreseeable future. Acoustics is one of the new environmental sciences that has just become a recognized and respected discipline within the past half century. The fruits of its expanding body of scientific knowledge, i.e., practical engineering applications, are also just beginning to be taken seriously on a widespread basis by architects, engineers, and planners in the solutions of acoustical problems in and around buildings.

Although there are a few exceptions, adequate courses in acoustics have been notably lacking in schools of architecture and engineering. A number of us, including Professor M. David Egan, were fortunate in learning about acoustics at the Massachusetts Institute of Technology through the pioneering and inspirational teaching of Professors Robert Newman, Leo Beranek, and Richard Bolt. Some of their former students, like Professor Egan, have gone on to spread the word through not only design applications of their knowledge but also through lecturing and teaching efforts.

Professor Egan has identified the pressing need for a textbook which would cut through to the core of the information needed to understand acoustics problems and to develop practical solutions. The book should hardly be called a textbook in the traditional sense, since the verbal descriptions are few and the major emphasis is on graphic displays of concepts as well as of engineering data and problem-solving techniques. This unique approach, as he has found with his students at the Clemson University College of Architecture and elsewhere, appeals to most students of architecture and engineering (of all ages) who need comprehensive yet encapsulated treatments of the environmental sciences. For these students, the real need is for a textbook which emphasizes concepts and the relative importance of the subject matter in the total environmental system for which the designer/decision-maker must be responsive.

This book is an important contribution to the better understanding of building acoustics problems in a growing multidisciplinary design environment.

WILLIAM J. CAVANAUGH
Fellow, Acoustical Society of America
Natick, Massachusetts

Preface

This book presents the basics of building acoustics in a graphical format. This kind of approach which emphasizes concept sketches also includes useful data for design and step-by-step example problem solutions. The book is directed to the student who wishes to learn the concepts of architectural acoustics as quickly as possible. It is important, therefore, that the reader carefully peruse all notes on the concept sketches. The sketches are not supplements to the text but in a very real sense are the text.

The graphical approach also should facilitate understanding of important technical concepts for those practicing professionals who have limited time available to digest lengthy verbal descriptions. In addition, the tables of engineering data, integrated within the graphical sequence, will be useful for solving actual problems.

Work on the book started during the author's tenure at Tulane University. The encouragement and support for this project by the late Professor John W. Lawrence, Dean of the School of Architecture, is deeply appreciated.

Thanks are due to Professor Robert B. Newman (Harvard University and M.I.T.), Mr. Parker W. Hirtle, and other friends and former colleagues at Bolt, Beranek and Newman, Inc., Cambridge, Mass., for their help in broadening the author's knowledge of acoustics.

Mr. William J. Cavanaugh, Consultant in Acoustics, Natick, Mass., and Mr. W. Ranger Farrell, Ranger Farrell and Assoc., Irvington-on-Hudson, New York, carefully reviewed the manuscript. Their many helpful suggestions are gratefully acknowledged.

M. David Egan, P.E.

Introduction

The architect must deal primarily with the human senses of sight, touch and hearing when designing a building's environment. Knowledge of the concepts of architectural acoustics, the study of everything which concerns hearing in or around buildings, will enable the designer to properly integrate the requirements of sound control with the requirements of other building functions. Every space that man occupies possesses an acoustical environment. Perhaps a carport doesn't have any acoustical problems in itself, but almost anything more complicated presents some kind of inherent acoustical problems that must be recognized and solved by the designer. (Even carports sometimes present acoustical problems since car noise can disturb occupants of adjacent bedrooms or neighboring buildings!) The goal of architectural acoustics is to make the environment best serve the functions intended, such as work, relaxation or sleep. All architects can benefit from a study of architectural acoustics for several important reasons.

The designer cannot equip himself to effectively solve acoustical problems from studies of oversimplified articles on acoustics in advertisers literature or in trade magazines. It is essential that the designer be prepared to recognize and solve acoustical problems from the very earliest design stages of a project. If not, such unfortunate situations can occur as the case involving a large eastern insurance company whose new office building had such inadequate interoffice privacy that in some instances every other office had to be vacated. This is inexcusable and would not have happened if proper studies had been made before the partitions and ceilings were specified. Every architect should be knowledgeable enough to protect himself and his clients from acoustical surprises in his completed buildings.

The majority of acoustical material supplier's technical handouts do not contain complete criteria needed for design. Their primary objective is often to present technical data in the most favorable manner. Certainly it is important to know a building material's general properties (e.g. fire endurance ratings, sound transmission class ratings, etc.), but it is vitally important to know what specific properties are needed to solve a particular problem. For example, it would be wasteful to use an A-grade sound isolating partition when the less costly B-grade would be adequate, and conversely it is essential to know the situations where only the A-grade will give satisfactory results. Consequently, the architect should be reasonably prepared to critically assess product advertising claims. The use of cleverly named products will not, unfortunately, reverse the immutable laws of physics governing the behavior of sound.

The architect should know when technical assistance is needed. Equipped with a knowledge of the concepts of architectural acoustics, the architect will be able to identify problems requiring assistance and engage in productive collaboration with all who are involved in the design process. Without a basic knowledge of acoustics, the architect-owner-consultant relationship could become somewhat of an example of Weizsäcker's observation: "Even if someone came today who knew the answers to all unsolved problems, we should not understand him if our own need had not already driven us to put the questions which he answers."

Many textbooks and technical papers on acoustics are research or engineering oriented. Unfortunately, when the architect is faced with using this material he quite understandably feels like a child who has been told to ride a bicycle by steering in a curve with a radius (R) proportional to the square of his velocity (v) divided by the angle of his imbalance (θ).

$$R \approx \frac{v^2}{\theta} \quad (\approx \text{ means varies as})$$

In other words, the architect cannot put such information to use and must resort to a trial and error process or simply ignore the problem. The disastrous situations that often result can be avoided by a detailed study and reasonable understanding of the concepts of architectural acoustics.

SOURCE-PATH-RECEIVER

Almost all acoustical situations have three common aspects—a source, a transmission path, and a receiver. Sometimes the source (humans, mechanical equipment, etc.) can be made louder or quieter, and the path (ground, building materials or air) can be made to transmit more or less sound. The receiver (usually humans, although sometimes animals or equipment) can also be influenced. For example, a building occupant can usually hear better if a distracting noise is removed. Focusing on only one aspect of a particular problem, however, will often result in overdesign or no solution at all.

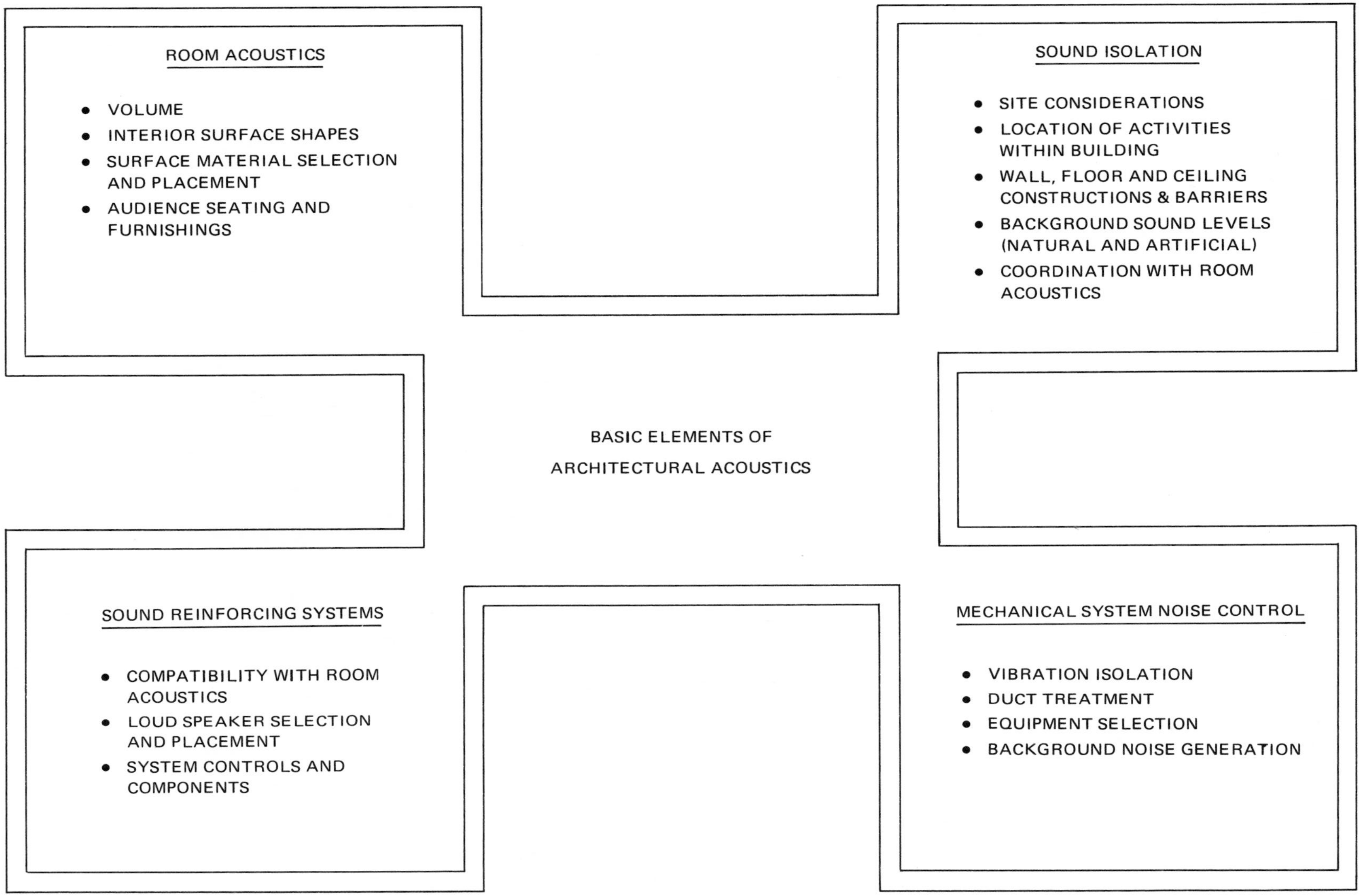

BASIC ELEMENTS OF
ARCHITECTURAL ACOUSTICS

ROOM ACOUSTICS

- VOLUME
- INTERIOR SURFACE SHAPES
- SURFACE MATERIAL SELECTION AND PLACEMENT
- AUDIENCE SEATING AND FURNISHINGS

SOUND ISOLATION

- SITE CONSIDERATIONS
- LOCATION OF ACTIVITIES WITHIN BUILDING
- WALL, FLOOR AND CEILING CONSTRUCTIONS & BARRIERS
- BACKGROUND SOUND LEVELS (NATURAL AND ARTIFICIAL)
- COORDINATION WITH ROOM ACOUSTICS

SOUND REINFORCING SYSTEMS

- COMPATIBILITY WITH ROOM ACOUSTICS
- LOUD SPEAKER SELECTION AND PLACEMENT
- SYSTEM CONTROLS AND COMPONENTS

MECHANICAL SYSTEM NOISE CONTROL

- VIBRATION ISOLATION
- DUCT TREATMENT
- EQUIPMENT SELECTION
- BACKGROUND NOISE GENERATION

Concepts in Architectural Acoustics

SECTION 1

Basic Theory

BASIC THEORY: Sound and Vibration

Sound is a vibration in an elastic medium (i.e., returns to normal state after force is removed) such as air shown below, building materials, and the earth. Sound energy progresses rapidly and can travel great distances. Each vibrating particle, however, moves only an infinitesimal amount to either side of its normal position. It "bumps" adjacent particles and imparts most of its motion and energy to them.

VIBRATION OF A PARTICLE IN AIR

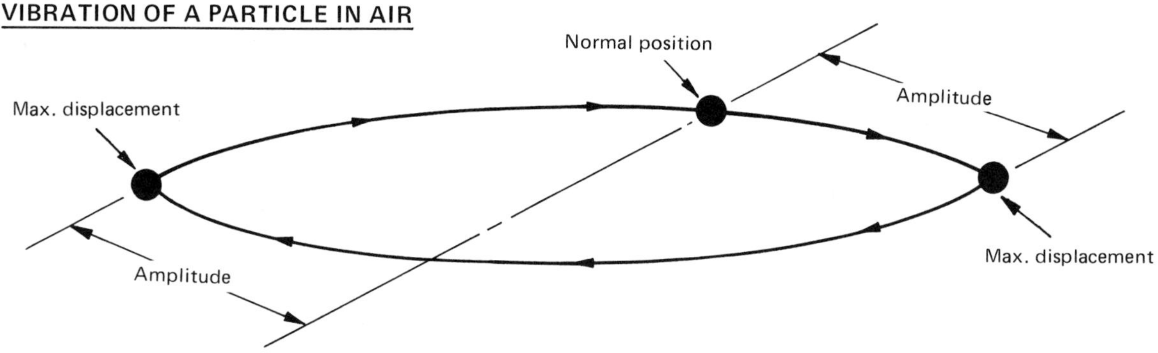

Cycle • Full circuit by the particle
Frequency • Number of complete cycles per second, called Hertz, abbrv. Hz
Amplitude • Maximum displacement of a particle to either side of its normal position during vibration

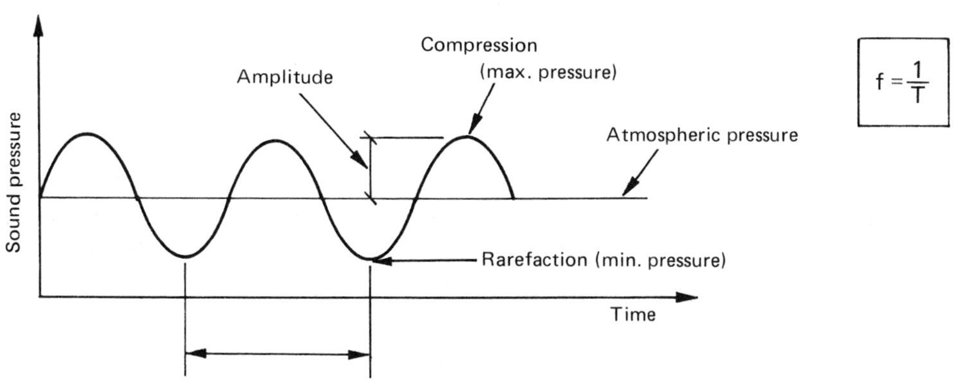

$$f = \frac{1}{T}$$

Period (T) in sec (time req. for one complete vibration or cycle)

Variation in pressure caused by a vibrating tuning fork (pure tone)

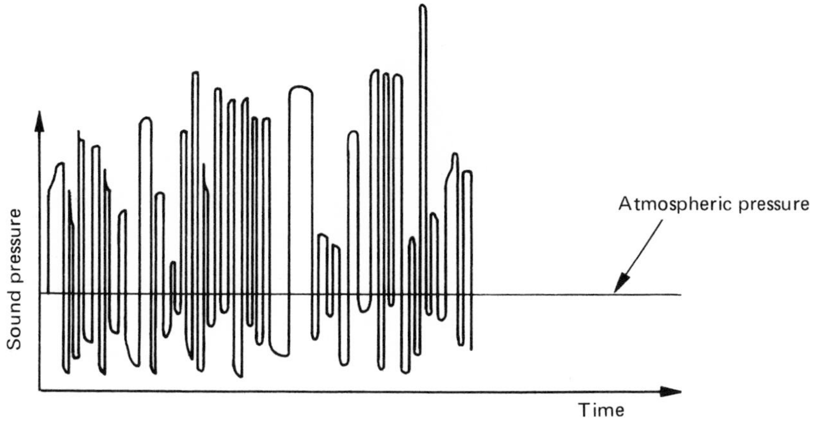

Variation in pressure caused by speech, music, or noise (complex sounds)

BASIC THEORY: Character of Sound

As sound passes through air, the "to-and-fro" motion of the particles alternately pushes together and draws apart adjacent particles, forming regions of rarefaction and compression. Wavelength is the distance between adjacent regions where identical conditions of particle displacement occur. Sound waves in air are analogous to the waves caused by a stone dropped in the water.

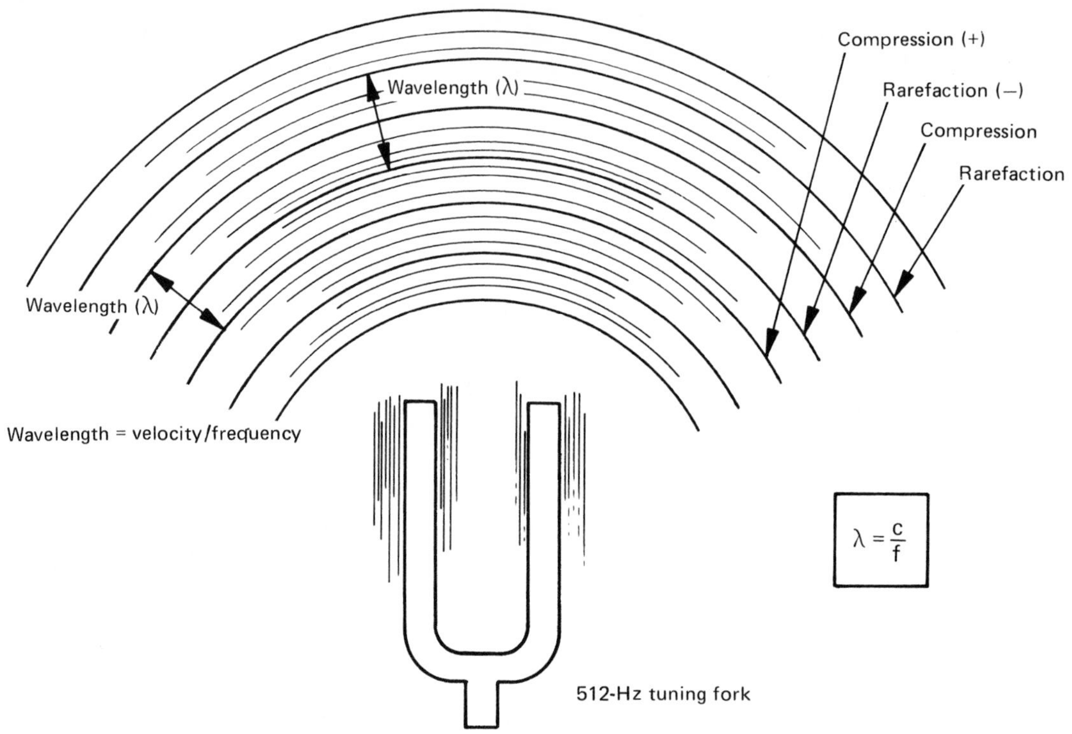

DESCRIPTION OF SOUND

Sound spectra are required for meaningful analysis of most sounds because almost all sounds are complex, consisting of numerous frequencies at different pressures. This requirement is similar to indoor climate control, where one cannot say 70° F is comfortable, without also specifying relative humidity, air motion, etc. Sound spectra define the magnitude of sound at many frequencies.

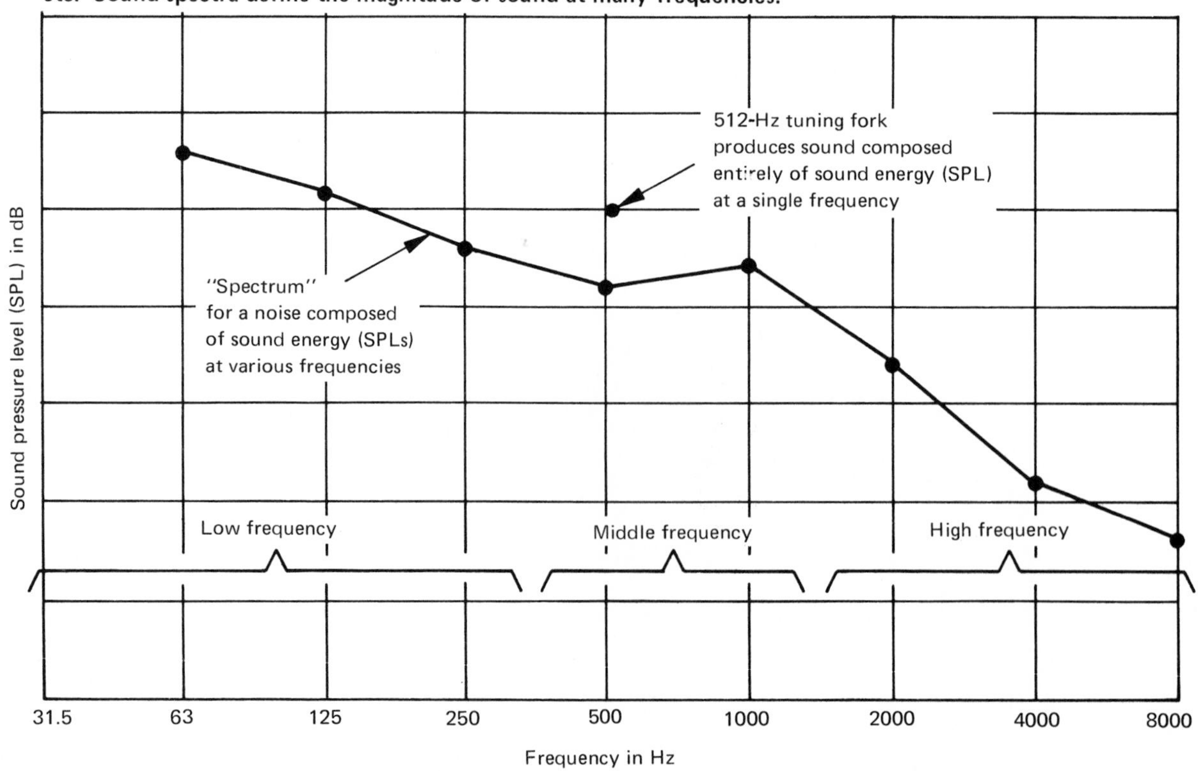

FREQUENCY OF SOUND

Frequency is the rate of repetition of a periodic event. Sound in air consists of a series of compressions and rarefactions due to air particles put into motion by a vibrating source. The frequency of a sound wave is determined by the number of times per second a given molecule of air vibrates about its neutral position. The greater the number of complete vibrations, the higher the frequency. The measure of frequency is the Hertz (Hz) or the number of cycles per second (cps) – the two are numerically equal and have the same meaning.

A healthy young person is capable of hearing sounds from about 20 to 20,000 Hz. The upper limit diminishes with increasing age – a condition called "presbycusis," which is generally more severe for men than women. Prolonged exposure to intense sound can also cause permanent hearing damage, and short-term exposure can cause temporary losses. Human speech contains energy from about 125 to 8000 Hz. Women's vocal cords are generally thinner and shorter than men's, which is the reason the female frequency of vibration (or pitch) is normally higher.

Most sound sources, except for pure tones (e.g., single-frequency sound from a tuning fork), contain energy over a wide range of frequencies. Consequently, in acoustics it is convenient to divide the audible frequency range into sections for meaningful analysis. For measurement and specification of sound one common division is into eight octave frequency bands identified by their center frequency as follows: 63, 125, 250, 500, 1000, 2000, 4000, and 8000 Hz. Just as with an octave on a piano keyboard, an octave in sound analysis represents the frequency interval between a given frequency (such as 125 Hz) and twice that frequency (250 Hz in this example).

Further divisions of the frequency range – half-octave bands, third-octave bands, etc. – are used for more detailed acoustical analyses. Sound-level meters can be used to measure energy at octave frequency bands by using electrical filters to eliminate the energy in the frequency region outside the band of interest. Note that the sound level covering the entire frequency range of the eight octave frequency bands listed above is referred to as the "overall" level.

VELOCITY OF SOUND

Sound travels at a velocity that depends primarily on the elasticity and density of the medium. In air, at normal temperature and atmospheric pressure, the velocity of sound is approximately 1130 feet per second (fps). This is extremely slow when compared with the velocity of light, which is 186,000 miles per second. It is, however, very fast when compared with air movement in conventional air-conditioning systems where the effects of air motion from room registers, etc., on sound velocity may be neglected. A very high air velocity of 2000 feet per minute (about 33 fps) is only about 3% of the velocity of sound. Consequently, in most air-conditioning system ducts sound goes upstream and downstream with equal ease!

The temperature effect on sound in buildings may also be neglected. For example, a significant 20°F change in room air temperature would cause only a 2% change in the velocity of sound.

BASIC THEORY: Frequency Range of Audible Sound

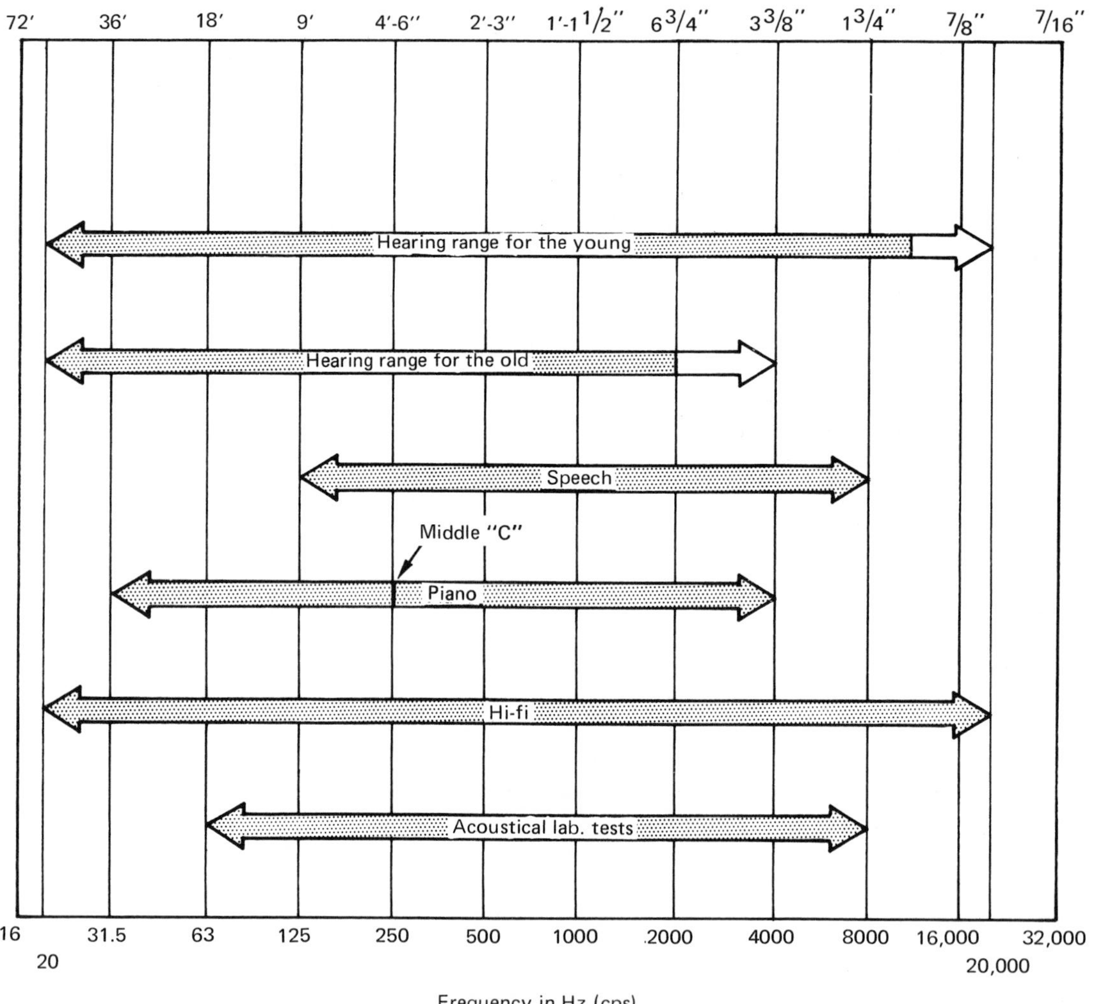

NOTE: Wavelengths (λ) are shown at top of graph above corresponding frequency.

BASIC THEORY: Inverse-square Law

Sound waves from a point source in the free field (e.g., outdoors with no obstructions) are virtually spherical and expand outward from the source, as shown below.

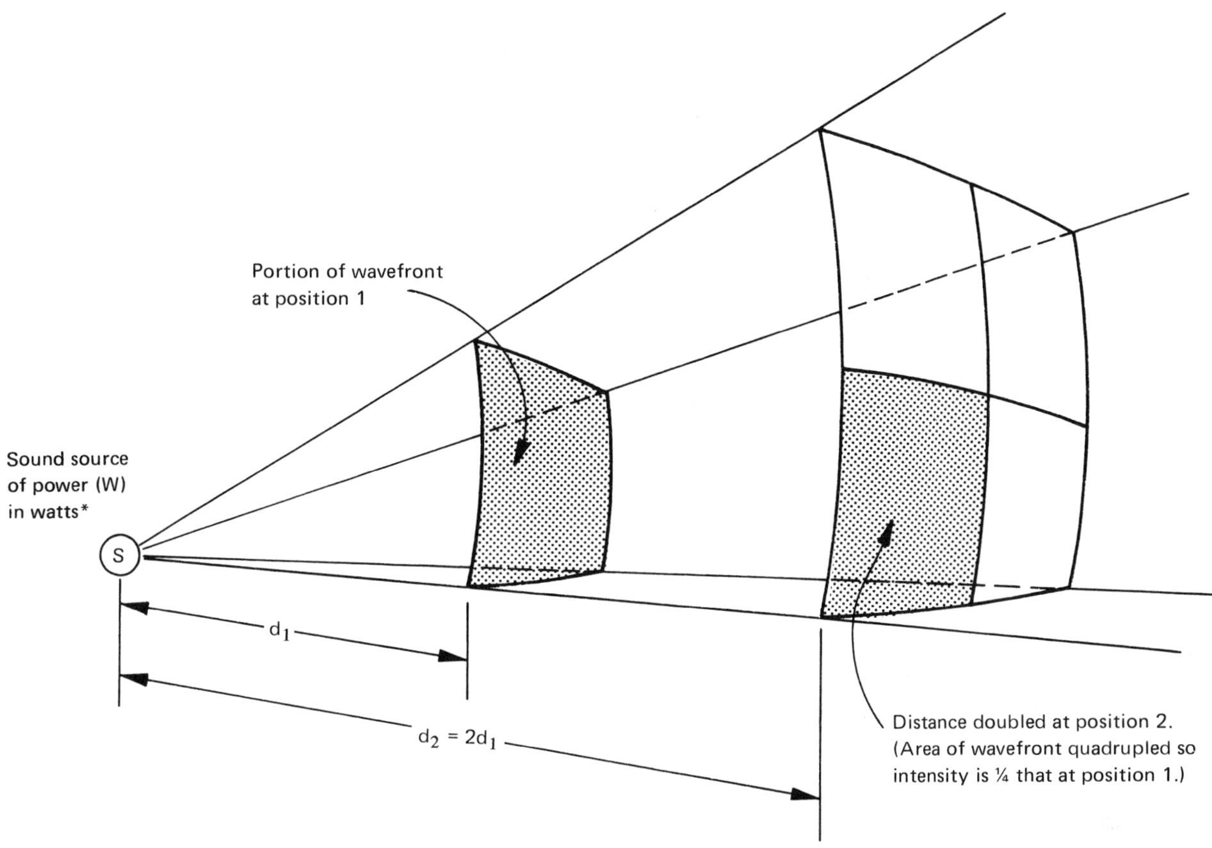

*Power is a basic quantity of acoustical and electrical energy. Although both are measured in watts they are different forms of energy and cause different responses. For instance, 10 watts of electrical energy through an incandescent bulb produces a very dim light, whereas 10 watts of acoustical energy through a loudspeaker can produce a very loud sound.

$$\text{Sound intensity} = \frac{\text{sound power}}{\text{total spherical area}}$$

Therefore: $I = \dfrac{W}{4\pi d^2}$ for sound in free field

where I = sound intensity, watt/cm^2
W = sound power, watt(s)
d = distance from sound source, cm

NOTE: If distance is in feet, multiply equations by 1/930 because 1 ft^2 = 930 cm^2.

At position 1.: $W = I_1 4\pi d_1^2$

At position 2.: $W = I_2 4\pi d_2^2$

Since W's are same, $I_1 4\pi d_1^2 = I_2 4\pi d_2^2$ and

$$\frac{I_1}{I_2} = \left(\frac{d_2}{d_1}\right)^2$$ ← The inverse-square law.

DECIBELS

According to the works of Ernst Weber and Gustav Fechner (nineteenth-century German scientists) nearly all sensations are proportional to the logarithm of the intensity of the source. In acoustics, therefore, to relate the intensity of sound to an intensity level corresponding to the human hearing experience, the "bel" unit, named in honor of Alexander Graham Bell, was introduced so that

$$IL = \log \frac{I}{I_0} \quad \text{(in bels)}$$

and

$$IL = 10 \log \frac{I}{I_0} \quad \text{(in decibels, or dB)}$$

where IL = sound intensity level, decibels (prefix "deci" indicating that logarithm is multiplied by 10)

I = sound intensity, watt per square centimeter (watt/cm^2)

I_0 = 10^{-16} watt/cm^2 (minimum sound intensity audible to average human ear)

The following page shows the decibel level of some common, easily recognized sounds. Note that the human hearing range from the threshold of audibility at 0 dB to the threshold of pain at 130 dB represents a tremendous intensity ratio of 10 trillion (10,000,000,000,000) to 1. This is such a wide range of hearing sensitivity that it may be hard to imagine. As an analogy consider a scale sensitive enough to weigh a human hair. If this same scale had the wide range of the human ears, it would also be able to weigh a 30-story high-rise building! Logarithms will enable us to handle easily the wide range of numbers in acoustics.

BASIC THEORY: Common Sounds in Decibels (dB)

Some common, easily recognized sounds are listed below in order of increasing sound intensity levels in decibels. The sound levels shown for occupied rooms are typical general activity levels only and do <u>not</u> represent criteria for design.

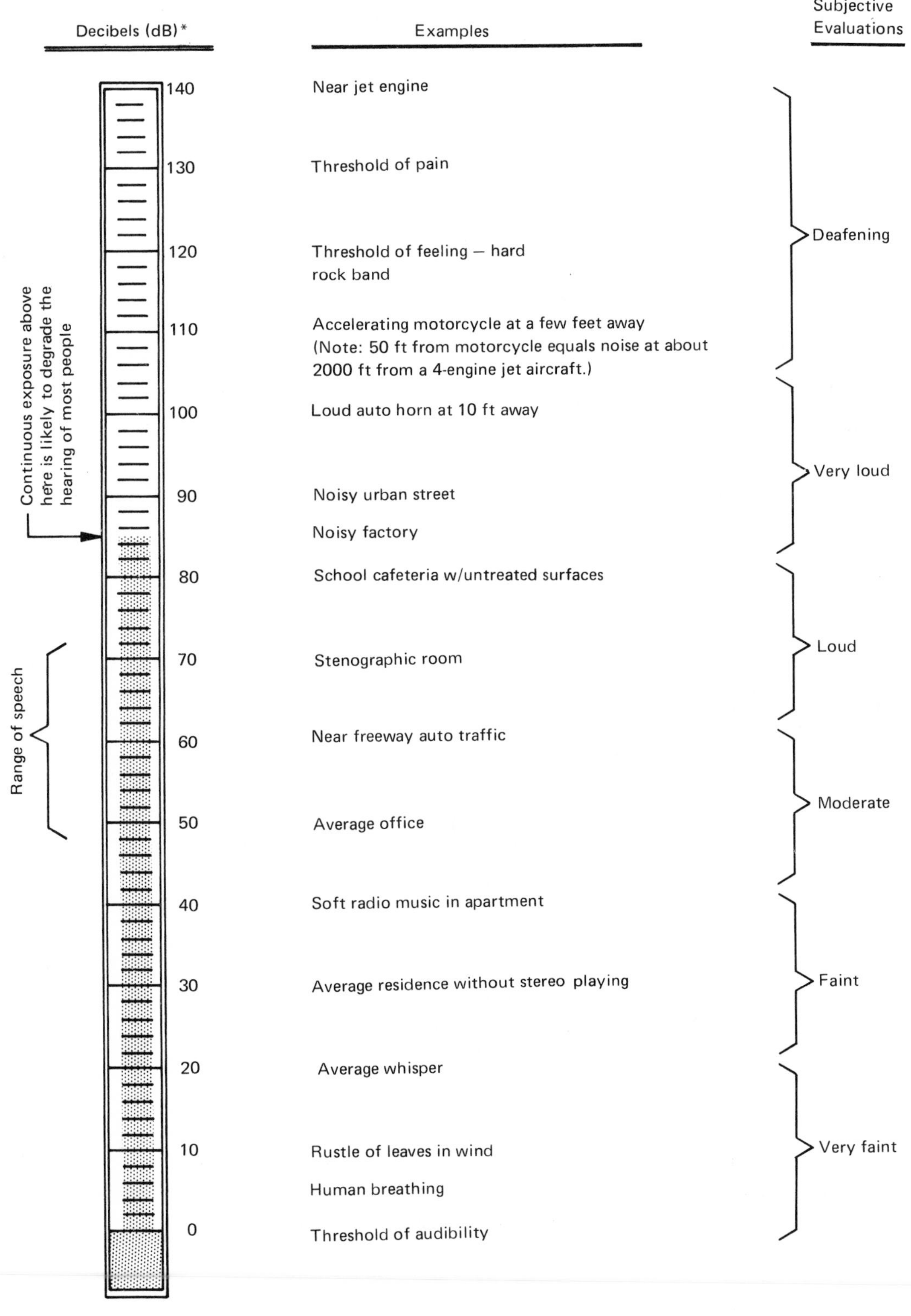

*dB are "average" values as measured on the A-scale of a sound-level meter.

The first step in finding the logarithm of a number is to express it as a digit from 1 to 9 multiplied by 10 to a power. A logarithm, in general, consists of two parts – the "characteristic," to which is added a decimal called the "mantissa." The characteristic is the power of 10, and the mantissa is found in "log" tables. In solving logarithms, remember that

$$10^5 = 100{,}000$$
$$10^4 = 10{,}000$$
$$10^3 = 1{,}000$$
$$10^2 = 100$$
$$10^1 = 10$$
$$10^0 \equiv 1 \qquad (\equiv \text{ means equal by definition to})$$
$$10^{-1} = 0.1$$
$$10^{-2} = 0.01$$
$$10^{-3} = 0.001$$

and when the decimal point is shifted to the left by n places, the number is to be multiplied by 10^n; when the decimal is shifted to the right by n places, the number is to be divided by 10^n. This may seem complex, but after reviewing a few examples it should be quite clear. ($\simeq$ means approximately equal to.)

$$4{,}820{,}000.0 = 4.82 \times 10^6 \simeq 5 \times 10^6$$

NOTE: Numbers ending in 0.5 and greater should be rounded up as shown above. For numbers ending below 0.5, merely drop the decimal part.

$$0.0000258 = 2.58 \times 10^{-5} \simeq 3 \times 10^{-5}$$
$$8{,}400{,}000{,}000.0 = 8.4 \times 10^9 \simeq 8 \times 10^9$$

The following condensed logarithm table can be used to quickly find the mantissa of numbers from 1 to 9. (See Appendix B for complete four-place log table and additional examples.)

A USEFUL LOG TABLE

Number	*Mantissa*
1	0
2	0.3
3	0.48
4	0.6
5	0.7
6	0.78
7	0.85
8	0.9
9	0.95

In almost all acoustical problems it is not necessary to work with small fractions of decibels. Use either the condensed log table above, or the four-place log table with your answers rounded to the nearest decibel.

Examples follow for both very large and very small numbers. Remember the first step is to arrange the number as a digit times 10 to a power.

$$\log(4{,}820{,}000.0) = \log(5 \times 10^6) = 6.7 = 6.7$$

└enter no. col. to find┘

$$\log(0.0000258) = \log(3 \times 10^{-5}) = -\log(1/3 \times 10^5)$$

$$= -\log(0.33 \times 10^5) = -\log(3 \times 10^4) = -4.48$$

$$\log(8{,}400{,}000{,}000.0) = \log(8 \times 10^9) = 9.9$$

ANTILOGARITHMS

The antilogarithm of a quantity, such as antilog x, is the number of which the quantity x is the logarithm. For example,

$$\text{antilog}(6.7) = 5 \times 10^6 = 5 \times 10^6$$

└enter mantissa col. to find┘

$$\text{antilog}(-4.48) = -3 \times 10^4 = 1/3 \times 10^{-4} = 0.33 \times 10^{-4} = 3 \times 10^{-5}$$

When a log's mantissa falls between values in the condensed log table, use the closest mantissa to find the corresponding number from 1 to 9.

BASIC PROPERTIES OF LOGS

$$\log xy = \log x + \log y$$

$$\log \frac{x}{y} = \log x - \log y$$

$$\log x^n = n \log x$$

$$\log 1 = 0$$

POWERS OF 10 REVIEW

Remember the symbol 10^3 is a shorthand notation for $10 \times 10 \times 10 = 1000$. That's all. Also, the product of two powers of the same number has an exponent equal to the sum of the exponents of the two powers:

$$10^2 \times 10^3 = (10 \times 10) \times (10 \times 10 \times 10) =$$
$$(10 \times 10 \times 10 \times 10 \times 10) = 10^5 = 10^{(2+3)}$$

Examples follow:

$$10^7 \times 10^5 = 10^{(7+5)} = 10^{12}$$

$$\frac{10^{-9}}{10^{-16}} = 10^{-9} \times 10^{+16} = 10^{(-9)+(16)} = 10^7$$

(Be *careful* of your signs.)

$$\frac{10^{-3}}{10^{-16}} = 10^{13}$$

You have now learned how to handle powers of 10. How easy it is!

BASIC THEORY: Example Problems

Given:

A car horn outdoors produces a sound intensity level of 90 dB at 10 ft away.

(1.) What is the horn's sound intensity at 10 ft?

(2.) What sound power does the horn radiate?

(3.) What will be the intensity level at 80 ft away from the horn?

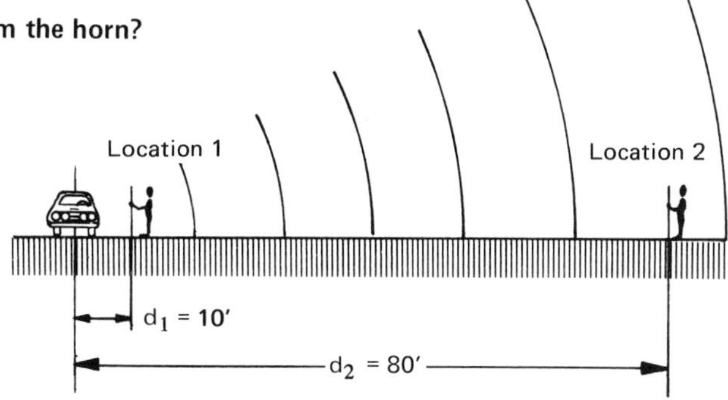

(1.) Intensity at 10 ft is found by

$$IL = 10 \log \frac{I}{10^{-16}}$$

$$90 = 10 \log \frac{I_1}{10^{-16}}$$

$$9 = \log \frac{I_1}{10^{-16}}$$

$I_1 = 10^9 \times 10^{-16} = 10^{-7}$ watt/cm^2 at 10′ away

(2.) Sound power can be found by the following formula if intensity is known at a given distance:

$$I = \frac{W}{4\pi d^2} \cdot \frac{1}{930}$$

and by cross-multiplication we get $W = 4\pi d^2 \times 930 \times 1$

By substitution $W = 4 \times 3.14 \times 10^2 \times 930 \times 10^{-7}$ because $I_1 = 10^{-7}$ watt/cm^2 at $d_1 = 10'$.

$W = 0.12$ watt

(3.) Intensity level at 80 ft away by inverse-square law:

$$\frac{I_1}{I_2} = \left(\frac{d_2}{d_1}\right)^2$$

$$\frac{10^{-7}}{I_2} = \left(\frac{80}{10}\right)^2 = 64$$

$$I_2 = \frac{1}{64} \times 10^{-7}$$

$I_2 = 1.56 \times 10^{-9}$ watt/cm^2 at 80′ away

and $$IL_2 = 10 \log \frac{I_2}{10^{-16}} = 10 \log \frac{1.56 \times 10^{-9}}{10^{-16}}$$

$IL_2 = 10 \log 1.56 \times 10^7 = 10\,(7.1931) = 72$ dB at 80′ away

which means a listener moving from location No. 1 to No. 2 at 80 ft away would observe a change in intensity level of 18 dB (i.e., 90 less 72 dB) – "very much quieter," as indicated by the table on the following page. Note, however, that the car horn at 72 dB would still be judged "loud" by most people.

CHANGES IN INTENSITY LEVEL

The following table is only an approximation of human sensitivity to changes in sound intensity level, because intensity isn't registered directly at the ear but is transferred to the brain, where it is recorded by its loudness. This makes our hearing highly individualized. Sensitivity to noise also depends on the frequency content, time of occurrence, duration, etc. Nevertheless, the table is a reasonable guide to help explain increases or reductions in decibel values for many architectural acoustics situations.

Change in Intensity Level, dB	*Change in Apparent Loudness*
1	Almost imperceptible
3	Just perceptible
5	Clearly noticeable
10	Twice (or one-half) as loud
18	Very much louder (or quieter)

Change in intensity level (or noise reduction) can be found by

$$NR = 10 \log \frac{I_1}{I_2}$$

where NR = difference in sound intensity levels between any two conditions IL_1 and IL_2, dB

I_1 = sound intensity, watt/cm^2, under one condition

I_2 = sound intensity, watt/cm^2, under another condition

MUSIC MAN EXAMPLE

Suppose that the sound of one trombone is received at our ears at an intensity level of 60 dB. Then the intensity level of sound from 76 trombones is found as follows:

One trombone:

$$IL_1 = 10 \log \frac{I_1}{I_0}$$

$$60 = 10 \log \frac{I_1}{10^{-16}}$$

$$6 = \log \frac{I_1}{10^{-16}}$$

$$I_1 = 10^6 \times 10^{-16} = 10^{-10} \text{ watt/cm}^2$$

76 trombones: If we assume that all the trombones play with identical frequency content, we can then combine their intensities. Therefore,

$$I_2 = 76I_1$$

$$IL_2 = 10 \log \frac{76I_1}{10^{-16}}$$

$$IL_2 = 10 \log \frac{76 \times 10^{-10}}{10^{-16}} = 10 \log 7.6 \times 10^7$$

$$IL_2 = 10(7.8808) = 79 \text{ dB}$$

which is not as much louder as one might expect. In fact, it would take 10 million trombones to reach the threshold of pain at 130 dB (although our threshold of discomfort or disgust might be reached at a much lower level). A good composer is aware that a large number of instruments playing the same melody may not produce a large sound impression. He uses large numbers of instruments to achieve a special coloring or blending in the overall sound from the individual instruments. For example, one solo violin by its position and frequency spectrum may dominate the sound of the orchestra.

BASIC THEORY: Decibel Addition

Since decibels are logarithmic values, they cannot be combined by normal algebraic addition. For example, 60 dB plus 60 dB does not equal 120 dB, but instead only 63 dB, as indicated by the table below.

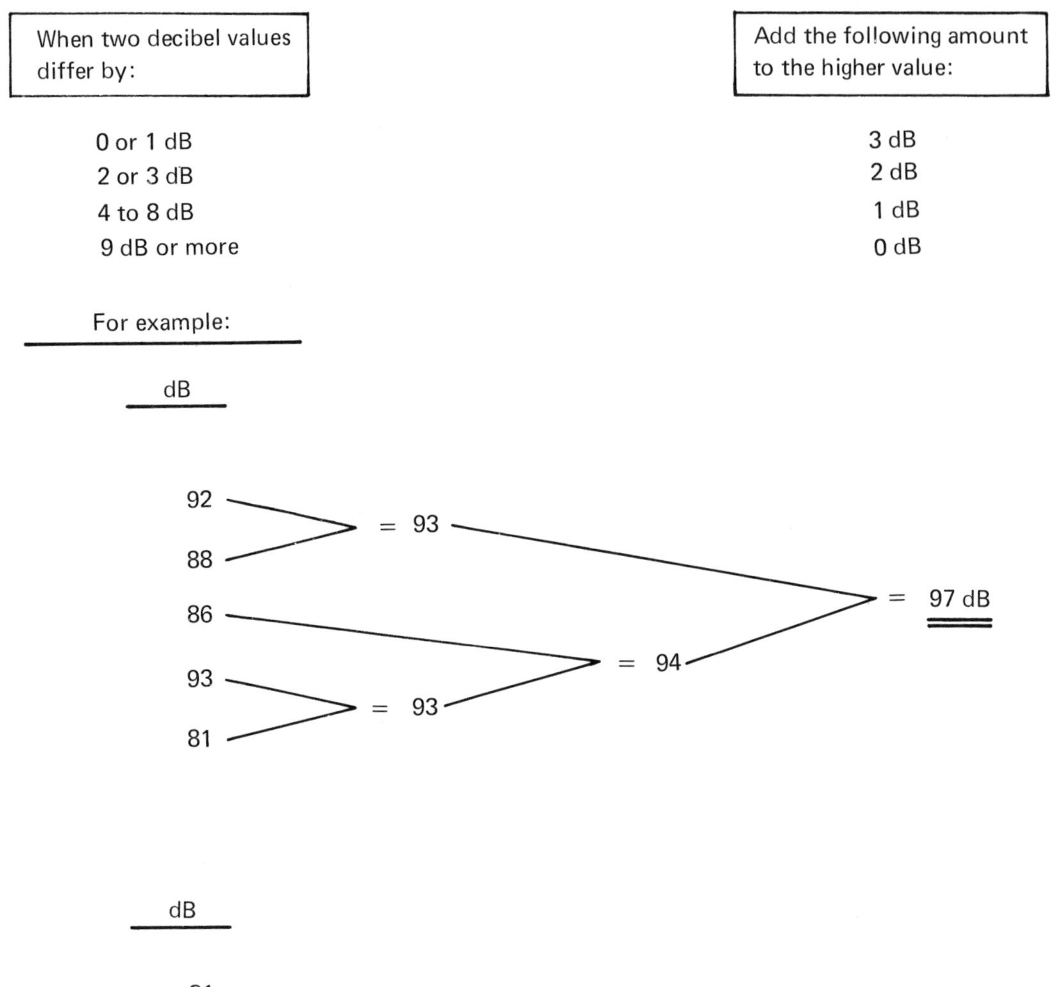

When two decibel values differ by:	Add the following amount to the higher value:
0 or 1 dB	3 dB
2 or 3 dB	2 dB
4 to 8 dB	1 dB
9 dB or more	0 dB

dB

81
86
= 87
= 91
88
92
93
= 96
= 97 dB

When combining decibels, be careful to sum only values from the same octave frequency band. Also, using different orders of addition may give results that differ by 1 dB, which is normally not too significant. To achieve precision, however, combine decibels logarithmically and use the tables to check your answer.

NOTE: **If a number of equal decibel values are to be combined, add 10 log n to the decibel value (where n is the number of equal decibel values). For example, if n = 76 trombones at an IL of 60 dB each, then total IL = 60 + 10 log 76 = 60 + 10 (1.8808) = 79 dB.**

BASIC THEORY: Sound Spectra showing the sound intensity levels in dB of four common noises in eight octave-band frequencies

Notice that the overall sound levels shown in brackets for the electric shaver and hair dryer are almost identical. However, they would not sound the same to a listener because their sound energy-frequency compositions differ, as indicated by their spectra shown below. Overall or single-number sound levels should be viewed with caution because human ears do not respond in an average manner since they are sensitive to frequency as well as pressure or intensity levels. For example, the ear is most sensitive to high frequency sounds (e.g., hissing or whistling sounds), as indicated by the curves on page 19.

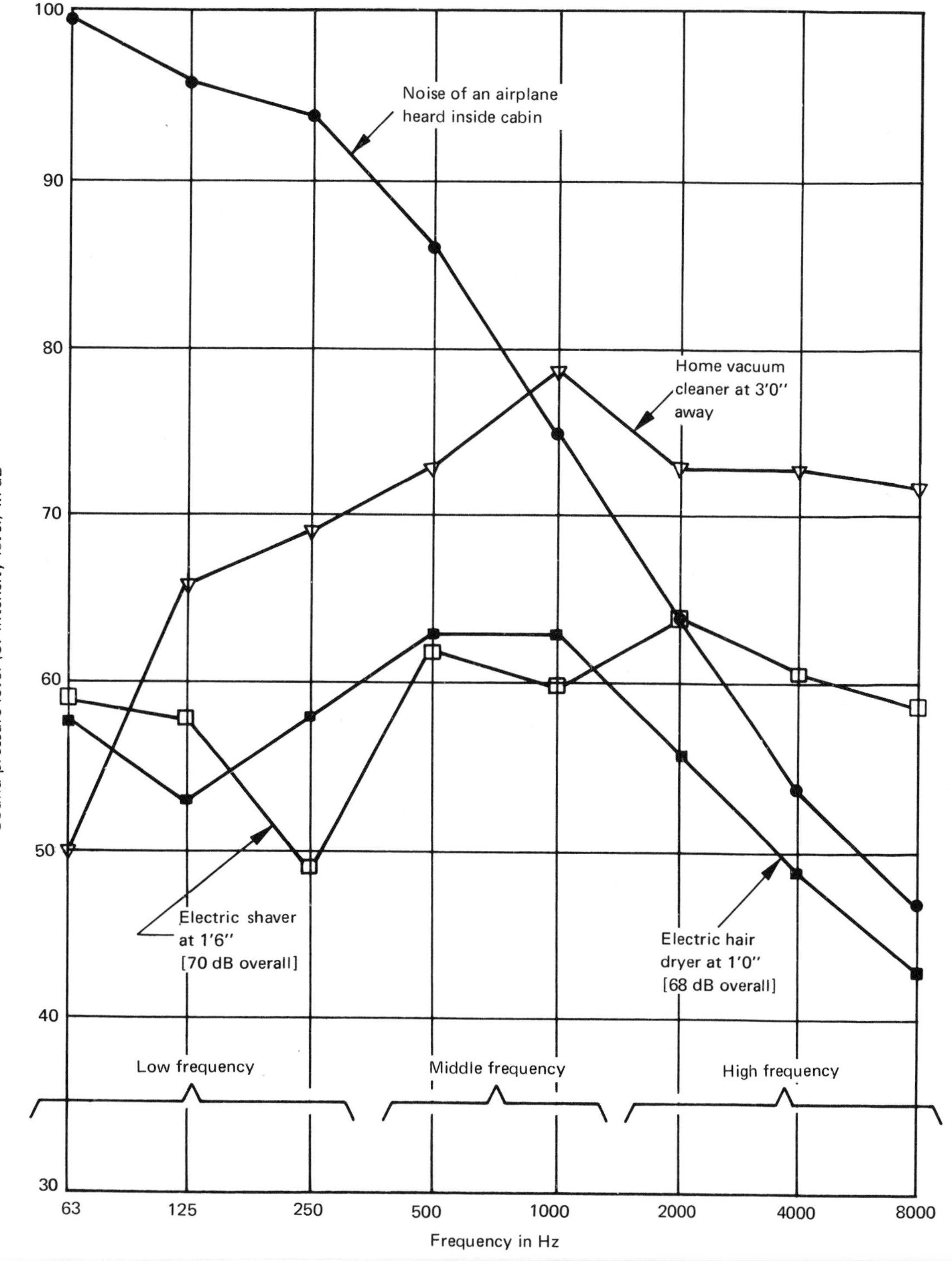

NOTE: Sound pressure level (SPL) is more conveniently measured than intensity level (IL). However, in most architectural acoustics situations they can be considered equal. The reference value for SPL in this book is always 0.0002 dyne/cm^2 (decibel values are meaningless w/o a reference value!).

BASIC THEORY: Human Ear

The ear sketch and comments below describe, in general, how the ear works.

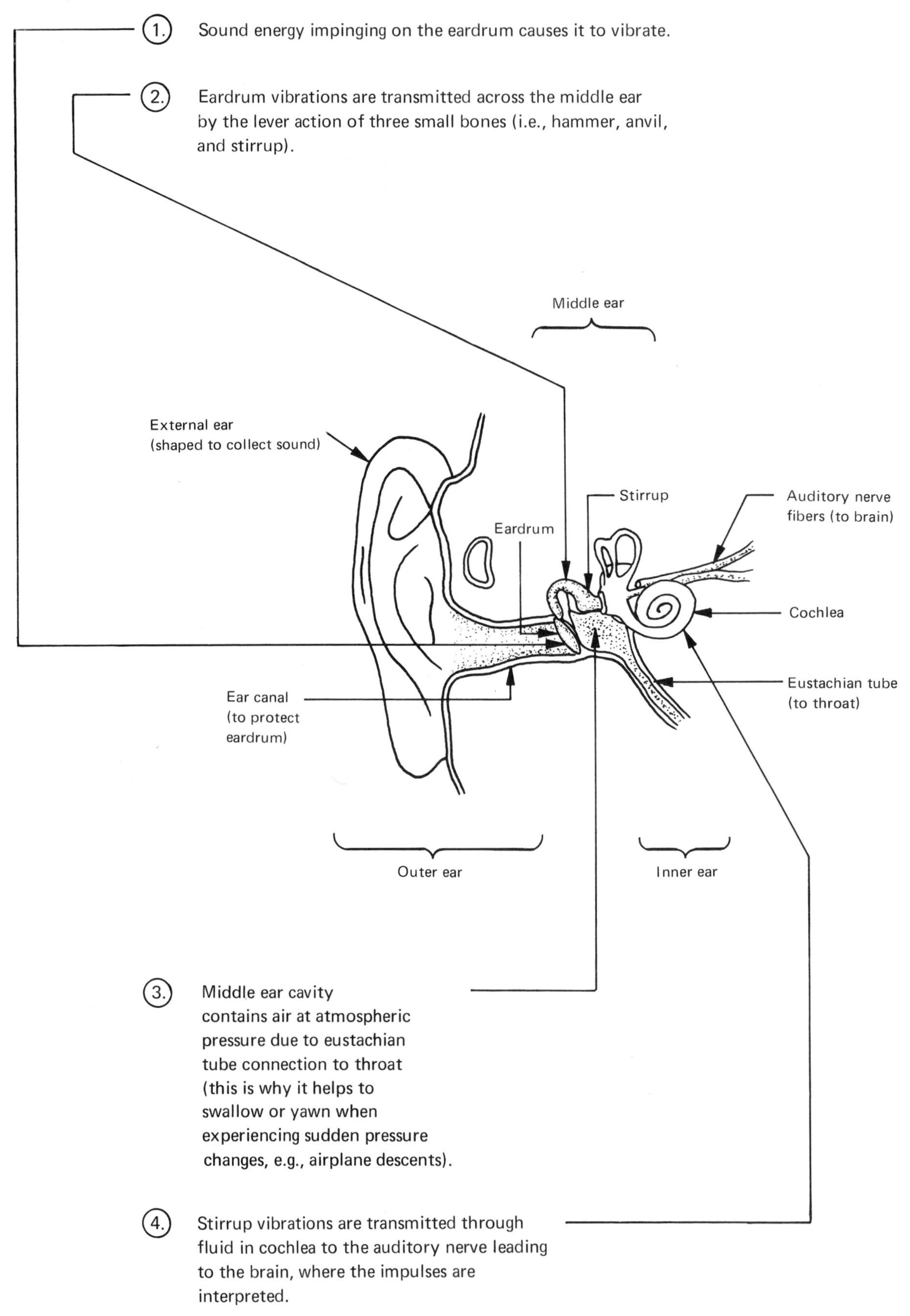

1. Sound energy impinging on the eardrum causes it to vibrate.

2. Eardrum vibrations are transmitted across the middle ear by the lever action of three small bones (i.e., hammer, anvil, and stirrup).

3. Middle ear cavity contains air at atmospheric pressure due to eustachian tube connection to throat (this is why it helps to swallow or yawn when experiencing sudden pressure changes, e.g., airplane descents).

4. Stirrup vibrations are transmitted through fluid in cochlea to the auditory nerve leading to the brain, where the impulses are interpreted.

BASIC THEORY: Equal Loudness Curves for octave-band noise—curves connect points of equal listener—judged loudness recorded in units similar to dB called "phons." (Based on hearing tests by Robinson and Whittle, reported in Acustica, vol. 14, 1964.)

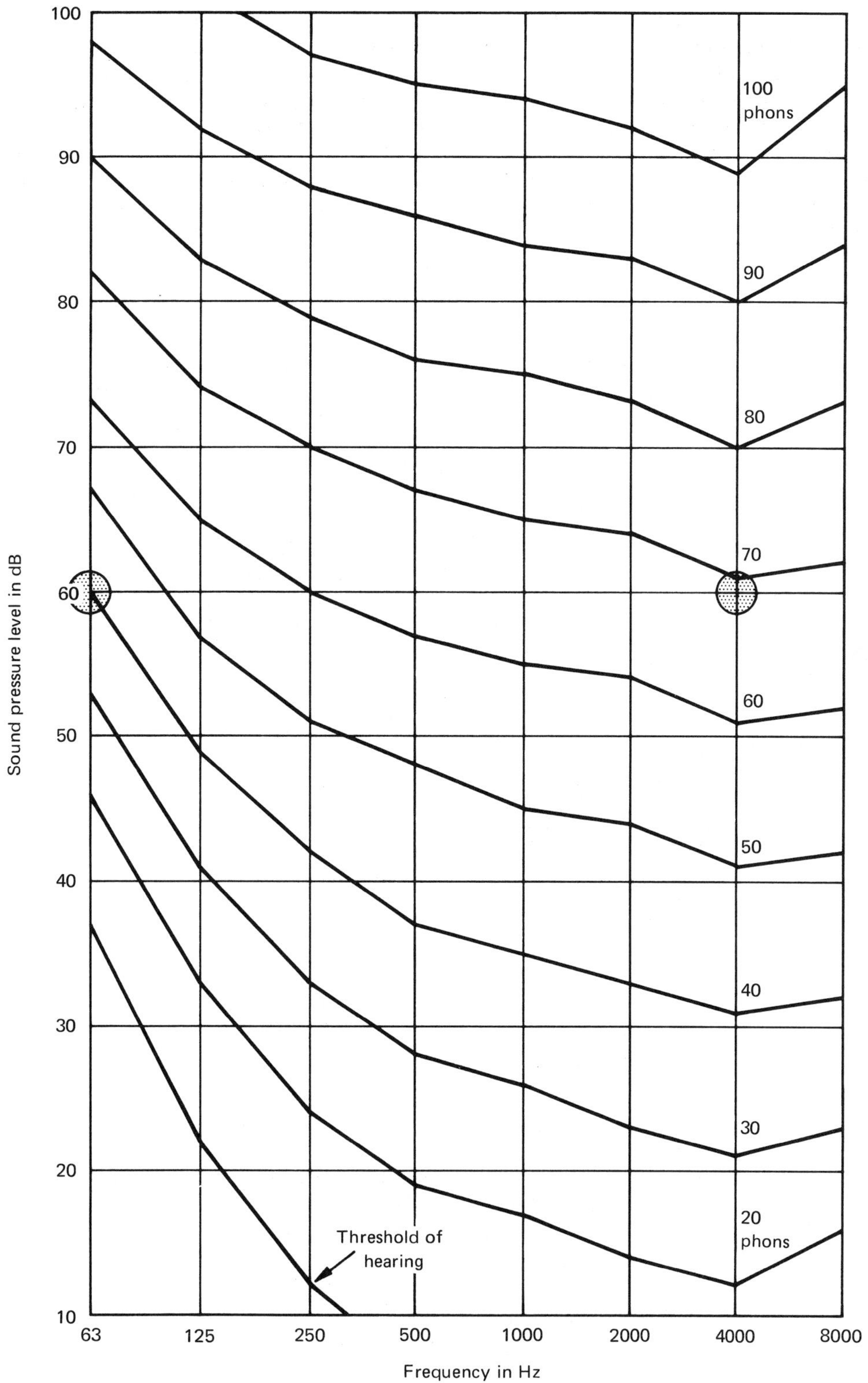

The curves numbered from 20 to 100 phons represent an equal loudness level judged by several test subjects. The curves show that human hearing is less sensitive to low frequency sound than to middle or high frequency sound of equal energy level. For example, 60 dB at 4000 Hz is judged louder, at nearly 70 phons, than the same 60 dB at 63 Hz, judged to be only 40 phons.

BASIC THEORY: Sound-level Meters and A-scale Decibels

A sound-level meter consists of a microphone which transforms the sound pressure variation in air into a corresponding electrical signal. This signal is then amplified internally and measured by appropriate electrical weighting networks with the results in decibels (dB) displayed on an indicating meter. The weighting networks tend to represent the frequency characteristics of the human ear by discriminating electronically in patterns similar to the equal loudness curves on page **19**. The standard networks are called A-scale, B-scale, and C-scale. The primary difference is that the A-scale tends to considerably neglect low frequency sound energy, the B-scale moderately, and the C-scale hardly at all. (Some meters also have a flat response with no frequency discrimination. This single-number result is called the "overall" decibel level.)

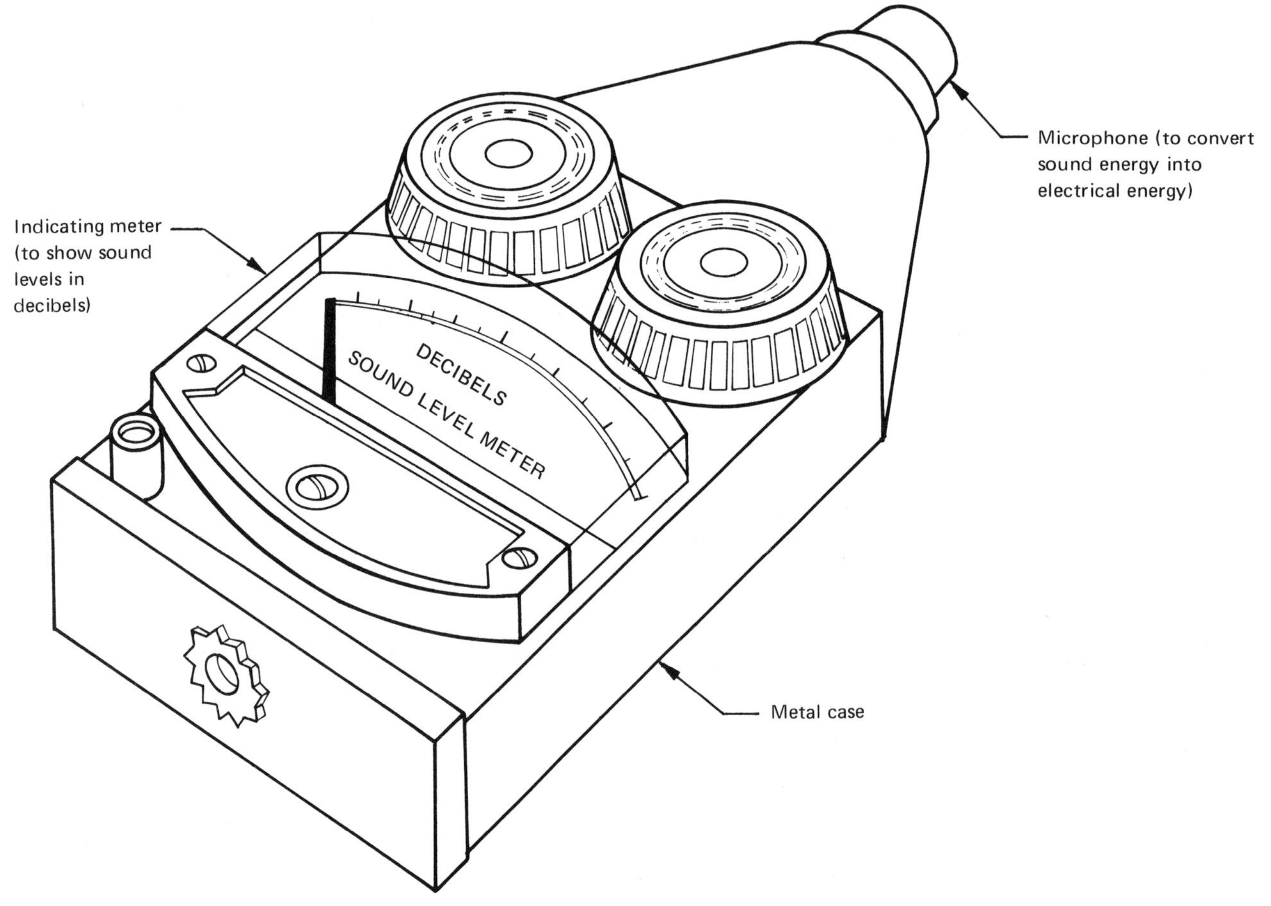

Typical hand-held sound-level meter

The A-scale weighting network can be used for some architectural work. For example, A-scale decibels (abbreviated "dBA") can be used to predict human reaction to some urban transportation noises. Shown below is the A-scale's discrimination against sound at different frequencies.

	Relative response (dB)							
	63 Hz	125 Hz	250 Hz	500 Hz	1000 Hz	2000 Hz	4000 Hz	8000 Hz
A-scale weighting network	−25	−15	−8	−3	0	+1	+1	−1

NOTE: For detailed analysis of any sound, more sophisticated equipment having octave-band frequency analysis capabilities should be used. A-scale decibels can be calculated from octave-band data by using the above relative response values and decibel addition. However, the reverse is not possible—dBA values cannot be converted to sound spectra!

TYPICAL NOISE-LEVEL DATA IN DECIBELS

Intermittent or peak values may exceed the decibel values in the table by 5 or more dB, depending on the type of source. For many practical problems, however, these typical source and general activity values can be used for design purposes if proper consideration is given to intermittent events which may exceed them.

TYPICAL SOUND PRESSURE LEVEL, dB

Source	63 Hz	125 Hz	250 Hz	500 Hz	1000 Hz	2000 Hz	4000 Hz	8000 Hz
1. Electric shaver at 1 1/2 ft	59	58	49	62	60	64	60	59
2. Vacuum cleaner at 3 ft	48	66	69	73	79	73	73	72
3. Garbage disposal at 2 ft	64	83	69	56	55	50	50	49
4. Washing machine	60	68	59	62	59	60	62	69
5. Window air-conditioning unit	64	64	65	56	53	48	44	37
6. Radio (average listening level)	. .	55	71	74	74	70	64	. .
7. TV at 10 ft	49	62	64	67	70	68	63	39
8. Stereo (teen-ager listening level)	60	72	83	82	82	80	75	60
9. Stereo (adult listening level)	56	65	75	72	70	66	64	48
10. Ringing telephone at 4 to 13 ft	. .	41	44	56	68	73	69	83
11. Ringing alarm clock at 4 to 9 ft	. .	46	48	55	62	62	70	80
12. Toilet, tank type, refill	50	55	53	54	57	56	57	52
13. Violin, at 5 ft (fortissimo playing)	. .	. .	91	91	87	83	79	66
14. Normal conversational speech	. .	60	75	78	75	65	55	38
15. Barking dogs	. .	. .	90	104	106	101	89	79
16. Amplified rock-and-roll music	. .	. .	. .	100	101	104	98	. .
17. Birds	. .	. .	. .	. .	. .	50	52	54
18. Trucks at 20 ft	89	86	81	77	73	70	67	64
19. Passenger cars at 20 ft	78	77	73	69	65	62	56	50
20. Jet planes at 2 to 3 miles in front of takeoff point	90	95	100	98	95	88	80	75
21. Motorcycles at 20 ft	92	96	93	89	79	73	70	63
22. Trains, pulling hard at 100 ft	95	102	94	90	86	87	83	79
23. Car horn at 15 ft	. .	. .	. .	92	95	90	80	60
24. Audio-visual rooms	85	89	92	90	89	87	85	80
25. Classrooms	60	66	72	77	74	68	60	50
26. Computer equipment rooms	78	75	73	78	80	78	74	70
27. Gymnasiums	72	78	84	89	86	80	72	64
28. Kitchens	86	85	79	78	77	72	65	57
29. Laboratory work spaces	65	70	73	75	72	69	65	61

TYPICAL SOUND PRESSURE LEVEL, dB								
Source	63 Hz	125 Hz	250 Hz	500 Hz	1000 Hz	2000 Hz	4000 Hz	8000 Hz
30. Libraries, reading	60	63	66	67	64	58	50	40
31. Mechanical equipment rooms	87	86	85	84	83	82	81	80
32. Music practice rooms	90	94	96	96	96	91	91	90
33. Reception and lobby areas	60	66	72	77	74	68	60	50

DECIBEL SCALES FOR SOUND INTENSITY, PRESSURE, AND POWER

	Sound Intensity Level	Sound Pressure Level	Sound Power Level
Abbreviation	IL	SPL	PWL
Express as	$10 \log \frac{I}{I_0}$	$20 \log \frac{P}{P_0}$	$10 \log \frac{W}{W_0}$
Units	I in watt/cm^2 IL in dB	P in dyne/cm^2 SPL in dB	W in watts PWL in dB
Weakest audible value (usually taken as the reference value)	$I_0 = 10^{-16}$ watt/cm^2	$P_0 = 2 \times 10^{-4}$ dyne/cm^2 (or 0.0002 microbar)	$W_0 = 10^{-12}$ watt
At weakest audible value	IL = 0 dB	SPL = 0 dB	PWL = 0 dB
Strongest tolerable value	$I = 10^{-3}$ watt/cm^2	$P = 2 \times 10^2$ dyne/cm^2	••
At strongest tolerable value	IL = 130 dB	SPL = 120 dB	••

SPL may be taken as equal to IL in most architectural acoustics problems. PWL expresses the amount of sound or noise power that is radiated by a given source, regardless of the space into which the source is placed. PWL is therefore somewhat analogous to the lumen rating of lamps in lighting. IL and SPL are somewhat analogous to room illumination levels. In other words, PWL is used to rate a sound source independently of its environment, while IL and SPL depend on the acoustical characteristics of the space in which the sound is heard.

SECTION 2

Sound Absorption

SOUND ABSORPTION: Sound Decay Outdoors and Within Enclosures

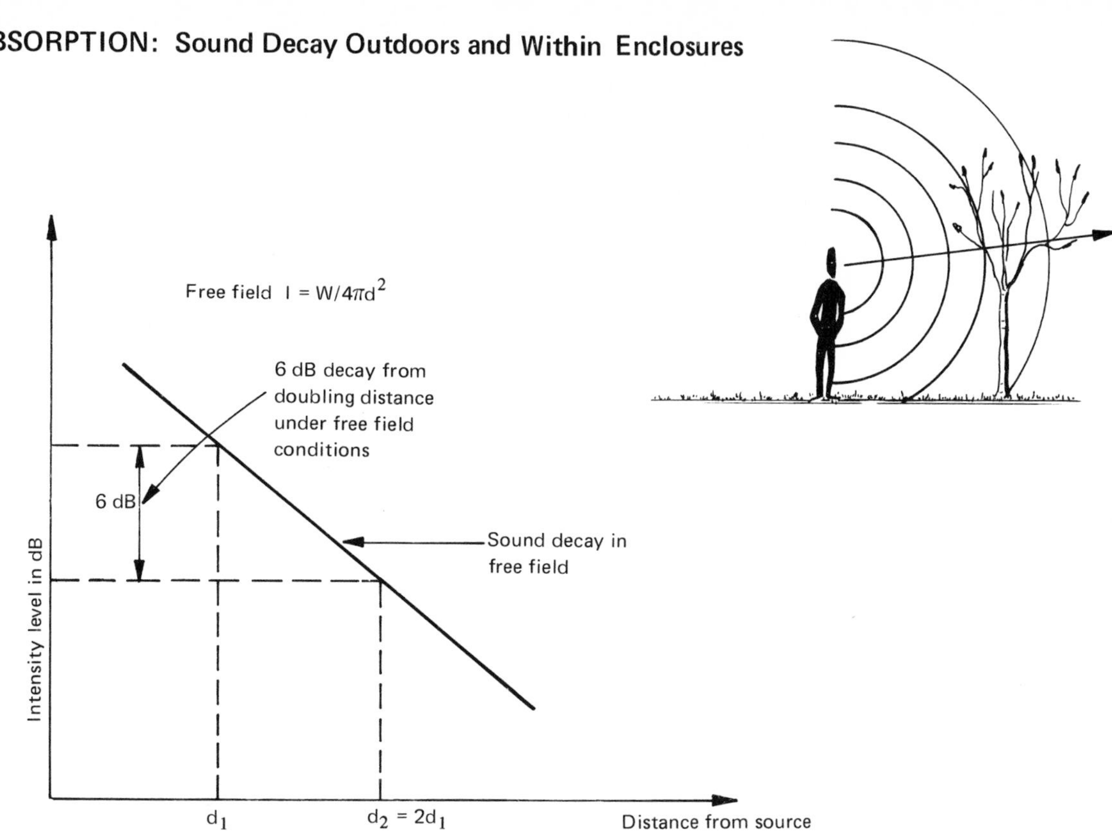

Sound decay outdoors in free field

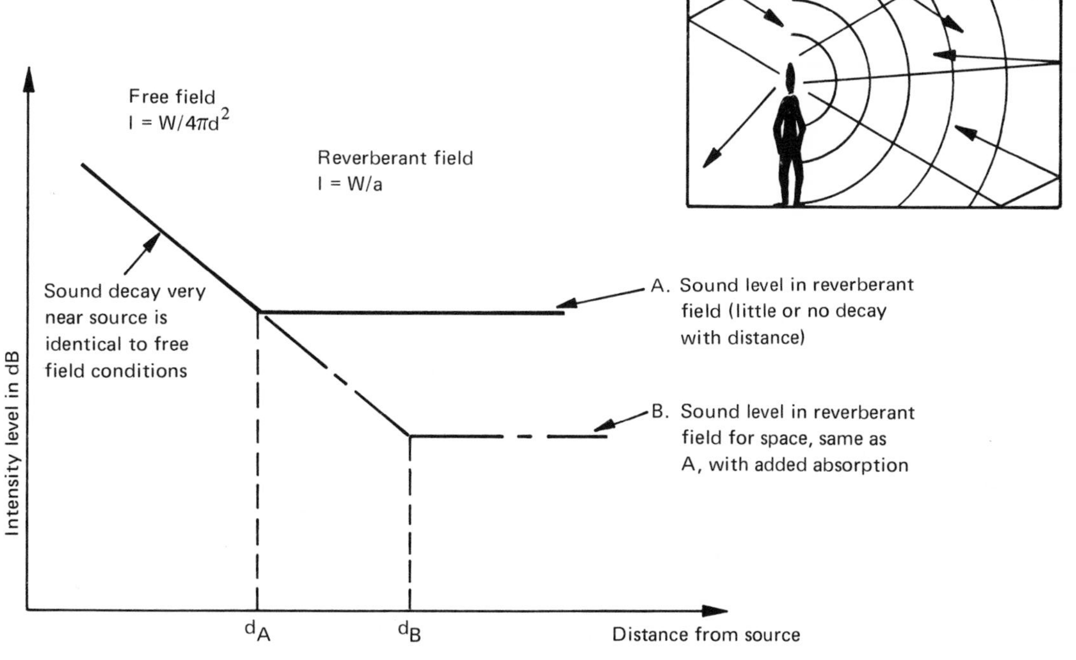

NOTE: Beyond distance $d = \sqrt{a/4\pi}$ from source, usually 1 to 4 ft, the intensity level is relatively constant and depends largely on total room absorption.

Sound decay indoors in reverberant field

SOUND ABSORPTION: Effect of Adding Sound-absorbing Material to a Room

In the room with no acoustical treatment, the reader hears direct sound from the TV as well as reflected sound from the ceiling, floor, and walls. The TV viewer, on the other hand, hears primarily direct sound. If sound-absorbing material is added to the room, the reader will hear considerably less reflected sound. Consequently, the sound level in his part of the room will be reduced. The sound level near the TV, however, is due mainly to direct sound, which remains unchanged.

SYMBOLS SHOWING DIRECTION OF SOUND WAVES

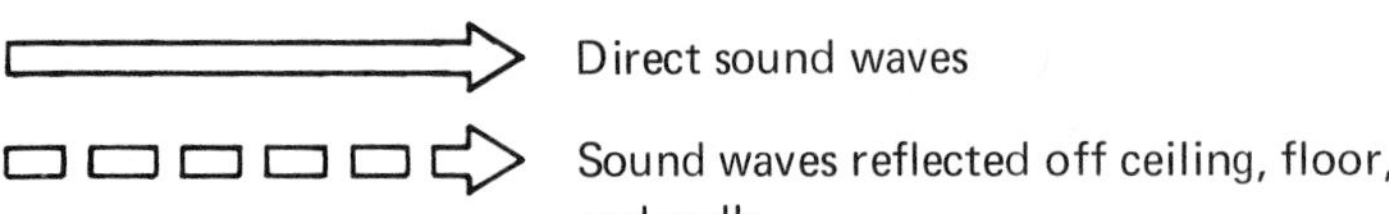

- Room with no acoustical treatment

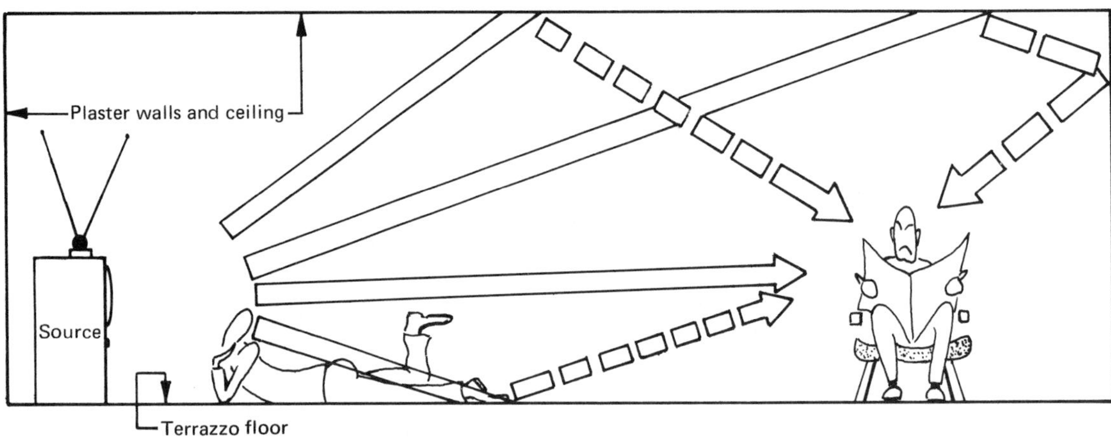

- Room with sound-absorbing material added

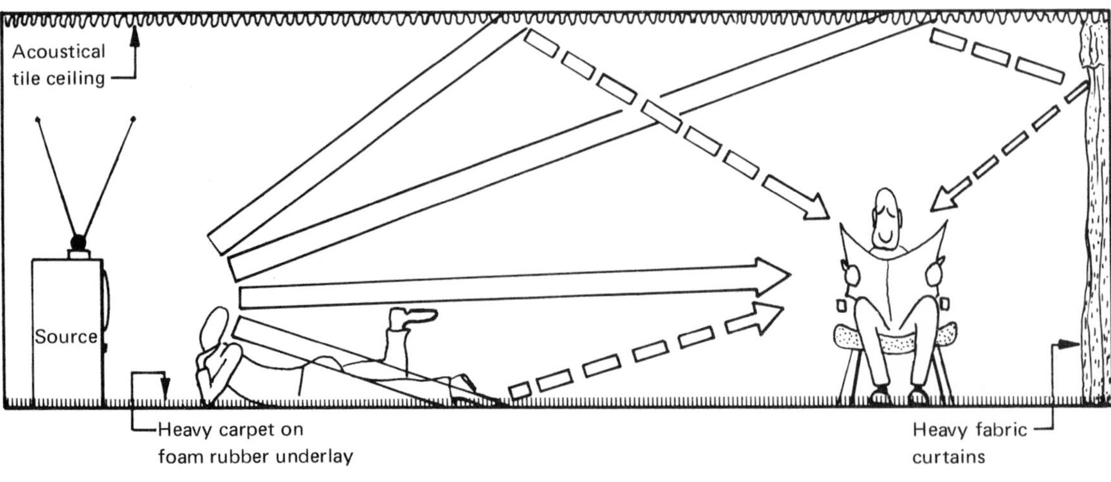

SOUND ABSORPTION: Effect of Room Surface Sound-absorbing Treatment

The addition of ceiling sound absorption to a 20′ by 20′ by 10′ high room reduces the sound level by 10 dB in the reverberant field, as shown below. However, close to the sound source the reduction is only about 3 dB. If the ceiling and all four walls are treated with sound-absorbing material, the sound level in the reverberant field drops an additional 6 dB, but the sound level near the source, in the free field, remains unchanged.

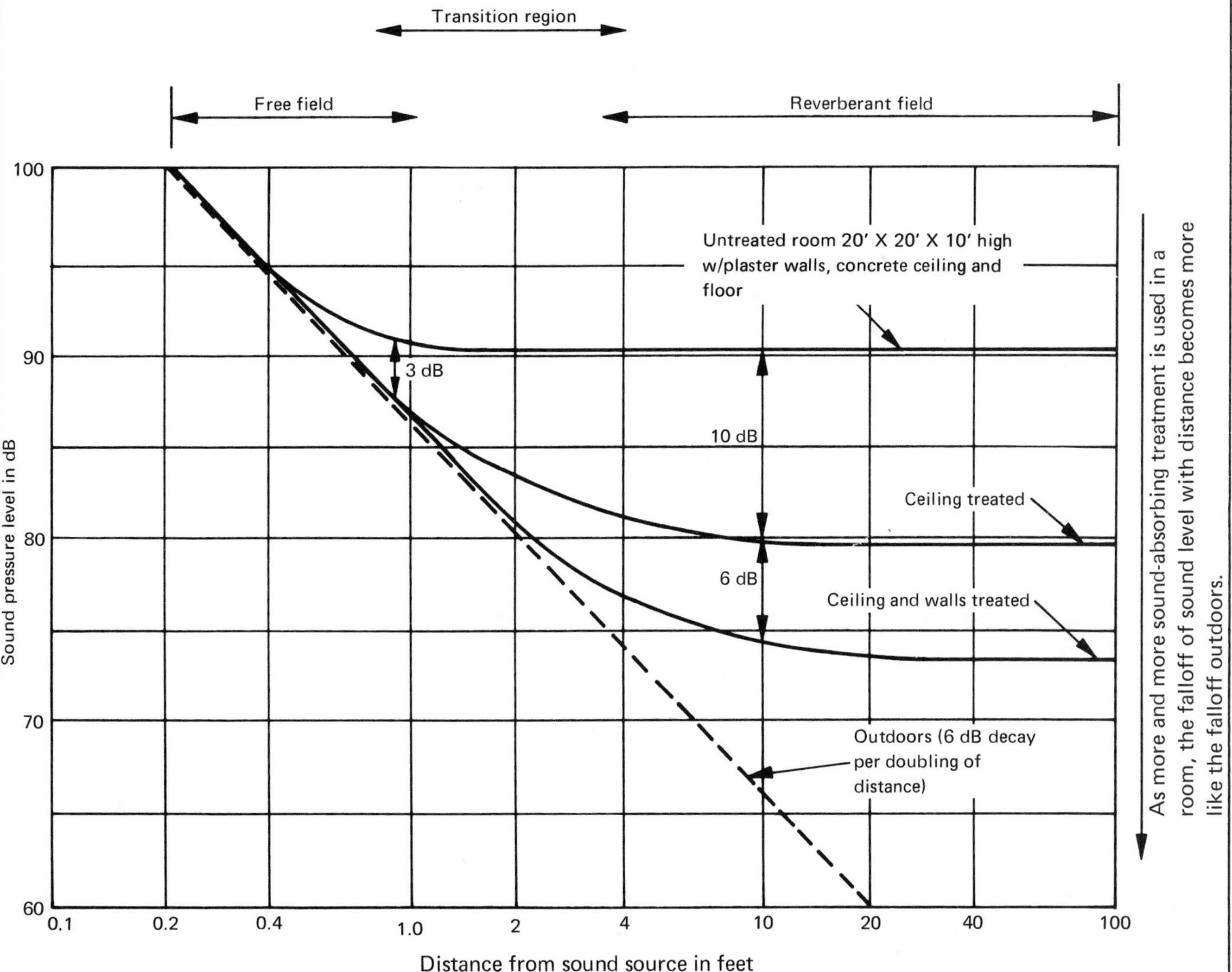

Ref.: A. C. Pietrasanta, Fundamentals of Noise Control, Noise Control, January, 1955, p. 13.

SOUND ABSORPTION: Measurements for Absorption

The effectiveness of any material as a sound absorber can be expressed by its absorption coefficient α. This coefficient describes the fraction of the incident sound energy that a material absorbs. Theoretically, it can vary from 0 (no sound absorption) to 1.0 (all incident sound absorbed). Coefficients are derived from laboratory tests or calculated from measurements in finished rooms.

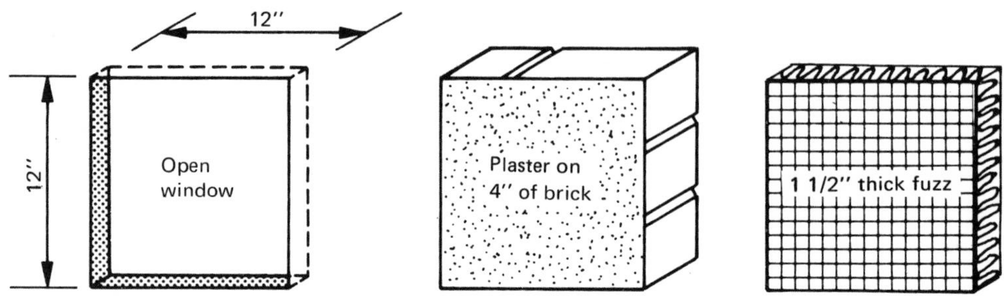

At 500 Hz for 1 sq ft of the above surface conditions:

Sound absorption	100%	2%	78%
Sabins (a)	1.0	0.02	0.78
Sound absorption coefficient α	1.0	0.02	0.78

- Total room absorption for a space is $a = \Sigma S\alpha$

  where a = total room absorption, sabins (the sum of all room surface areas times their respective sound absorption coefficients)

  S = surface area, ft^2

  α = sound absorption coefficient (dimensionless) at a given frequency

- Absorption coefficients for building materials normally vary from about 0.01 to 0.99.

  Materials having high sound absorption coefficients (usually meaning greater than about 0.20) are referred to as "sound-absorbing," whereas those with low coefficients are "sound-reflecting." In most situations, the effect of a difference in coefficients at a given frequency is shown by the following table:

Difference in Coefficient α	Effect for Most Situations*
0.05–0.10	Little
0.10–0.20	Significant
0.20 and above	Considerable

*NOTE: Exceptions are rooms used for acoustical lab. tests. For example, "reverberant" rooms used to measure α's for materials must have highly reflective surfaces. Even very small differences in room surface α's are therefore important.

SOUND ABSORPTION: Effect of Thickness on Absorption Efficiency

Sound absorption by porous sound absorbers (called "fuzz") is predominantly the indirect conversion of sound energy into thermal energy. The sound energy loses its energy through frictional flow against the walls of the mazelike interstices. The amount of absorption is determined by the porous absorber's actual physical properties of (1) thickness, (2) density, (3) porosity, and (4) fiber orientation. The optimization of these properties is in the realm of the researcher and manufacturer. However, it is evident from the curves below that thickness has a significant effect on porous absorbers. Also, it is essential that a porous material's internal structure be composed of interconnected pores. For example, plastic and elastomeric foams which have closed, nonconnected voids provide little or no sound absorption. A simple test for a porous material is to blow smoke through it. If the material is thick and readily passes smoke under moderate pressure, it should be a good absorber.

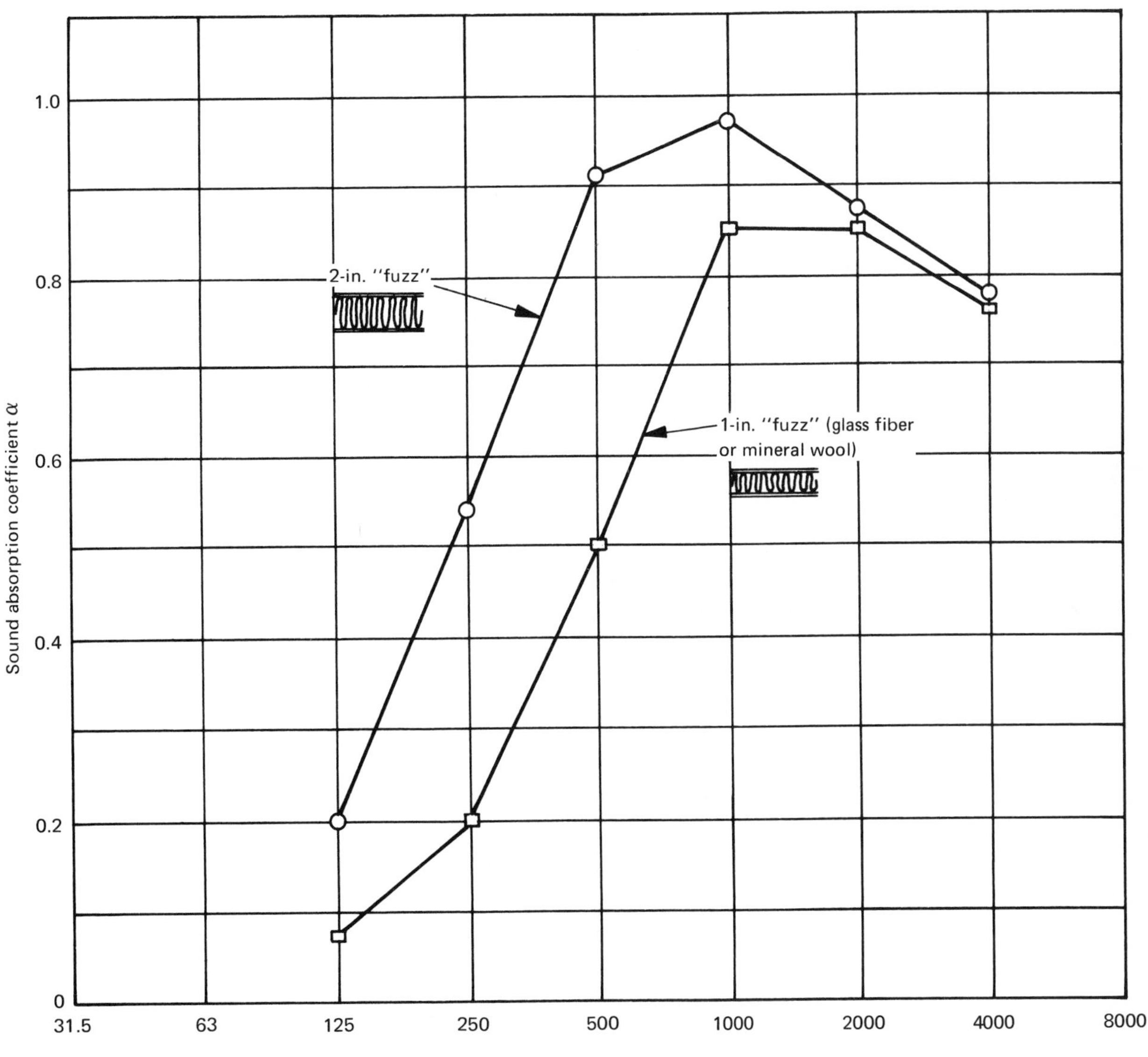

SOUND ABSORPTION: Relative Efficiency of Sound-absorbing Materials

NOTE: By careful design using porous materials and vibrating panels, uniform sound absorption with respect to frequency can be achieved. This is especially important in recording and radio/TV studios.

(1.) Basic types of sound-absorbing material

- Porous materials (convert sound energy to heat by friction)

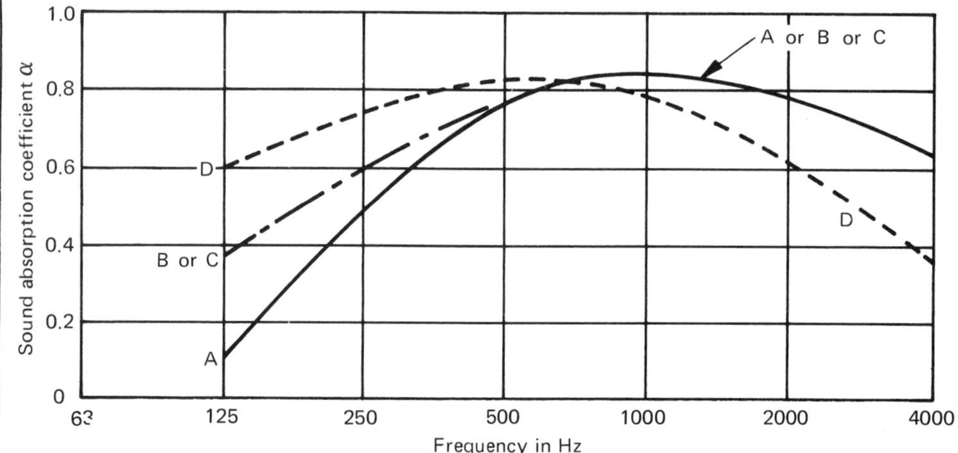

A Thin ½″ → 1½″

B Thick 1½″ → 4″

C Thin over air space

D Thick w/perf. facing

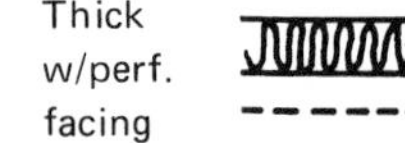

(2.) Specialized types of sound-absorbing material*

- Vibrating panels (convert sound energy to vibrational energy)

*Used to supplement porous absorption or to absorb specific low frequency sound.

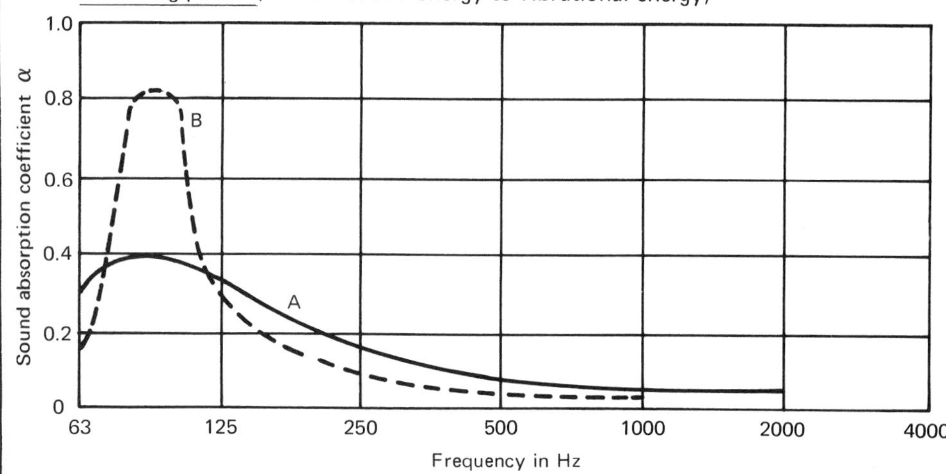

A With porous mat'l. in air space

B Without porous mat'l.

- Volume resonators (reduce sound energy by friction at opening and reduce energy within the cavity)

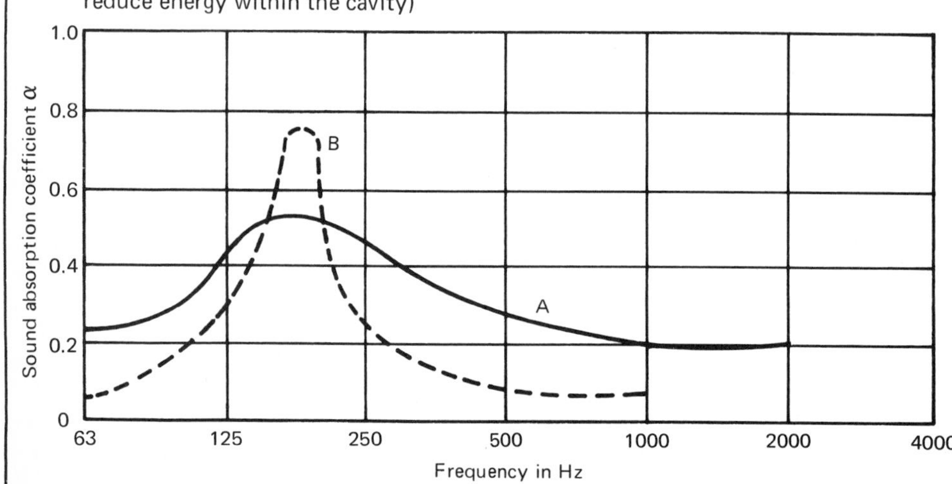

A With porous mat'l. in cavity

B Without porous mat'l.

SOUND ABSORPTION: Noise Reduction Coefficient

The noise reduction coefficient (NRC) is the arithmetical average of a material's sound absorption coefficients at 250, 500, 1000, and 2000 Hz carried to the nearest 0.05.

$$\text{NRC} = \frac{\alpha 250 + \alpha 500 + \alpha 1000 + \alpha 2000}{4}$$

- The NRC is intended as a single-number index of sound-absorbing efficiency. However, caution should be exercised when selecting a product based on its NRC. For example, the NRC does not include the α at 125 Hz, which can be an important factor where low frequency absorption is needed, such as in small music practice rooms. Also, because the NRC is an average number, two materials may have identical NRC's but very different absorption characteristics. Nevertheless, where low frequency absorption is not an important factor (such as in lobbies and general activity spaces), the NRC is an adequate index.

- Example—NRC calculation: Data from curve for carpet below.

$$\text{NRC} = \frac{0.20 + 0.35 + 0.45 + 0.55}{4} = \frac{1.55}{4} = 0.39 \quad \text{and} \quad \text{NRC} = 0.40 \text{ for carpet}$$

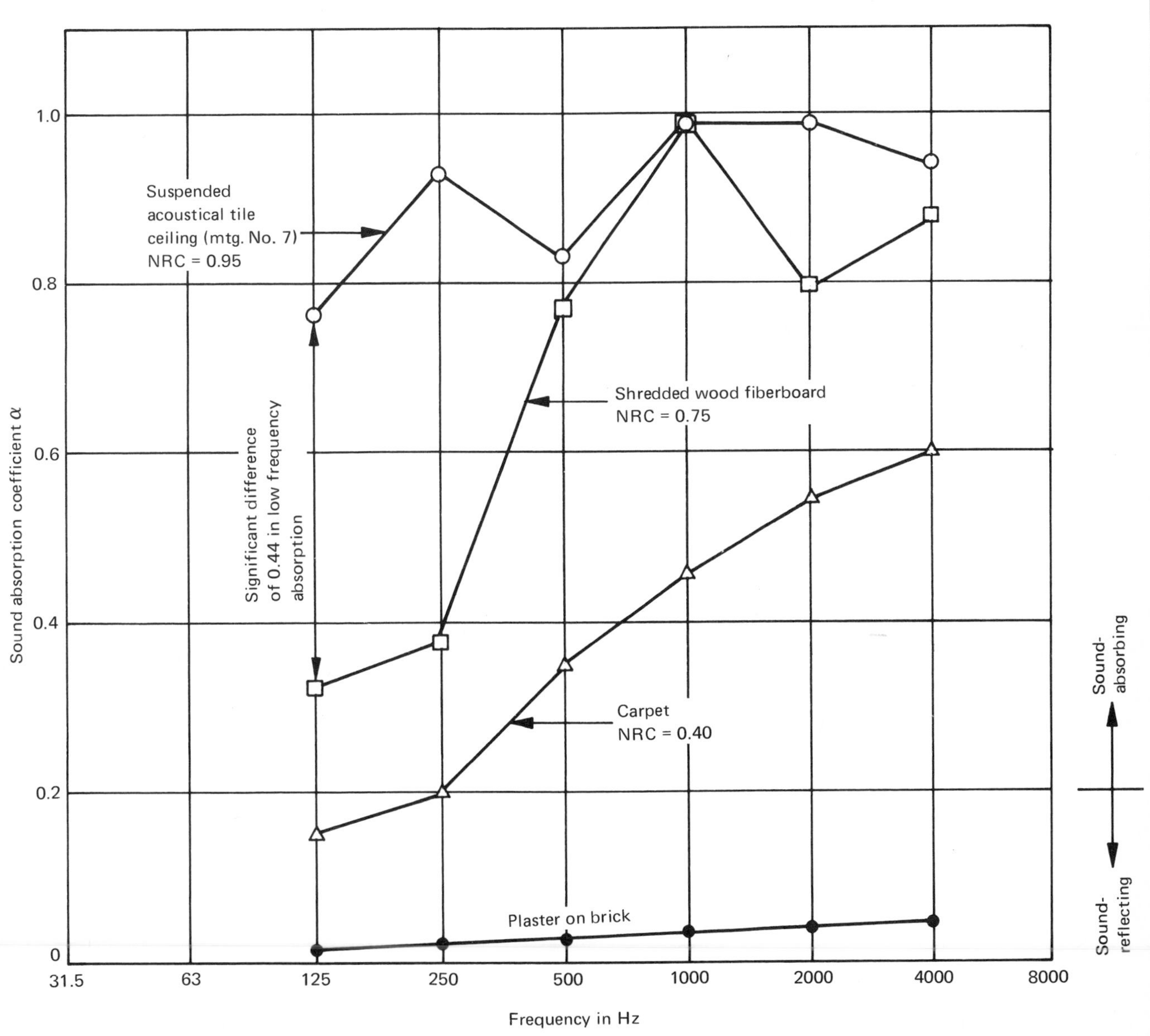

SOUND ABSORPTION DATA FOR COMMON BUILDING MATERIALS AND FURNISHINGS

Material	Sound Absorption Coefficient						NRC
	125 Hz	250 Hz	500 Hz	1000 Hz	2000 Hz	4000 Hz	Number*
Walls(1, 2, 5)							
Sound-reflecting:							
1. Brick, unglazed	0.03	0.03	0.03	0.04	0.05	0.07	0.05
2. Brick, unglazed and painted	0.01	0.01	0.02	0.02	0.02	0.03	0.00
3. Concrete block, painted	0.10	0.05	0.06	0.07	0.09	0.08	0.05
4. Cork on brick or concrete	0.02	0.03	0.03	0.03	0.03	0.02	0.05
5. Glass, heavy plate	0.18	0.06	0.04	0.03	0.02	0.02	0.05
6. Glass, typical window	0.35	0.25	0.18	0.12	0.07	0.04	0.15
7. Gypsum board, 1/2-in. paneling	0.29	0.10	0.05	0.04	0.07	0.09	0.05
8. Marble or glazed tile	0.01	0.01	0.01	0.01	0.02	0.02	0.00
9. Metal venetian blinds	0.06	0.05	0.07	0.15	0.13	0.17	0.10
10. Plaster, gypsum or lime, on brick	0.01	0.02	0.02	0.03	0.04	0.05	0.05
11. Plaster, gypsum or lime, on concrete block	0.12	0.09	0.07	0.05	0.05	0.04	0.05
12. Plaster, gypsum or lime, on lath	0.14	0.10	0.06	0.05	0.04	0.03	0.05
13. Plywood, 3/8-in. paneling	0.28	0.22	0.17	0.09	0.10	0.11	0.15
14. Wood, 1/4-in. paneling, with air space behind	0.42	0.21	0.10	0.08	0.06	0.06	0.10
Sound-absorbing:							
15. Concrete block, coarse	0.36	0.44	0.31	0.29	0.39	0.25	0.35
16. Cork, 1 in. with air space behind	0.14	0.25	0.40	0.25	0.34	0.21	0.30
17. Lightweight drapery, 10 oz/sq yd, flat on wall (note: sound-reflecting at most frequencies)	0.03	0.04	0.11	0.17	0.24	0.35	0.15
18. Mediumweight drapery, 14 oz/sq yd, draped to half area	0.07	0.31	0.49	0.75	0.70	0.60	0.55
19. Heavyweight drapery, 18 oz/sq yd, draped to half area	0.14	0.35	0.55	0.72	0.70	0.65	0.60
20. Fiberglas fabric curtain, 8 1/2 oz/sq yd, draped to half area	0.09	0.32	0.68	0.83	0.39	0.76	0.55
21. Shredded wood fiberboard, 2 in. thick on concrete (mounting No. 4)	0.32	0.37	0.77	0.99	0.79	0.88	0.75
22. Thick, porous sound-absorbing material with open facing	0.60	0.75	0.82	0.80	0.60	0.38	0.75
23. Carpet, heavy, on 5/8 in. perforated mineral fiber-board with air space behind	0.37	0.41	0.63	0.85	0.96	0.92	0.70

*NRC (noise reduction coefficient) is an average single-number rating of a material's sound absorption coefficients. It does not include the coefficients at 125 or 4000 Hz, which are often important information. Also, the name "noise reduction coefficient" is poorly chosen since noice reduction refers to the difference in sound intensity levels between any two conditions (or rooms). See page 31 for a discussion of the NRC's uses and limitations.

Material	Sound Absorption Coefficient						NRC
	125 Hz	250 Hz	500 Hz	1000 Hz	2000 Hz	4000 Hz	Number*
24. Wood, 1/2-in. paneling, perforated 3/16-in.-diameter holes, 11% open area, with 2 1/2-in. glass fiber in air space behind	0.40	0.90	0.80	0.50	0.40	0.30	0.65
Floors(5)							
Sound-reflecting:							
25. Concrete or terrazzo	0.01	0.01	0.02	0.02	0.02	0.02	0.00
26. Cork, rubber, linoleum, or asphalt tile on concrete	0.02	0.03	0.03	0.03	0.03	0.02	0.05
27. Marble or glazed tile	0.01	0.01	0.01	0.01	0.02	0.02	0.00
28. Wood	0.15	0.11	0.10	0.07	0.06	0.07	0.10
29. Wood parquet on concrete	0.04	0.04	0.07	0.06	0.06	0.07	0.05
Sound-absorbing:							
30. Carpet, heavy, on concrete	0.02	0.06	0.14	0.37	0.60	0.65	0.30
31. Carpet, heavy, on foam rubber	0.08	0.24	0.57	0.69	0.71	0.73	0.55
32. Carpet, heavy, with impermeable latex backing on foam rubber	0.08	0.27	0.39	0.34	0.48	0.63	0.35
33. Indoor-outdoor carpet	0.01	0.05	0.10	0.20	0.45	0.65	0.20
Ceilings(5)							
Sound-reflecting:							
34. Concrete	0.01	0.01	0.02	0.02	0.02	0.02	0.00
35. Gypsum board, 1/2-in. thick	0.29	0.10	0.05	0.04	0.07	0.09	0.05
36. Plaster, gypsum or lime, on lath	0.14	0.10	0.06	0.05	0.04	0.03	0.05
37. Plywood, 3/8-in. thick	0.28	0.22	0.17	0.09	0.10	0.11	0.15
Sound-absorbing:†							
38. Suspended acoustical tile, 3/4-in. thick (mounting No. 7)	0.76	0.93	0.83	0.99	0.99	0.94	0.95
39. Shredded wood fiberboard, 2 in. thick on lay-in grid (mounting No. 7)	0.59	0.51	0.53	0.73	0.88	0.74	0.65
40. Thin, porous sound-absorbing material, 3/4 in. thick (mounting No. 1)	0.10	0.60	0.80	0.82	0.78	0.60	0.75
41. Thick, porous sound-absorbing material, 2 in. thick (mounting No. 1), or thin material with air space behind (mounting No. 2)	0.38	0.60	0.78	0.80	0.78	0.70	0.75
42. Sprayed cellulose fibers, 1 in. thick on concrete (mounting No. 4)	0.08	0.29	0.75	0.98	0.93	0.76	0.75

*See p. 32

†Refer to manufacturers' catalogs or AIMA Performance Data Bulletin, which lists data for several hundred commercial sound-absorbing tile and panel materials. Data should be from up-to-date tests by acoustical laboratories per current ASTM procedures.

Material	Sound Absorption Coefficient 125 Hz	250 Hz	500 Hz	1000 Hz	2000 Hz	4000 Hz	NRC Number*
Seats and Audience(1, 3–5)‡							
43. Fabric well-upholstered seats, with perforated seat pans, unoccupied	0.19	0.37	0.56	0.67	0.61	0.59	
44. Leather-covered upholstered seats, unoccupied	0.44	0.54	0.60	0.62	0.58	0.50	
45. Audience, seated in upholstered seats §	0.39	0.57	0.80	0.94	0.92	0.87	
46. Chairs, metal or wood seats, each, unoccupied	0.15	0.19	0.22	0.39	0.38	0.30	
47. Students, informally dressed, seated in tablet-arm chairs	0.30	0.41	0.49	0.84	0.87	0.84	
Openings(5)¶							
48. Deep balcony, upholstered seats			0.50–1.00				
49. Grilles, mechanical system air			0.15–0.50				
50. Stage			0.25–0.75				
Miscellaneous(2, 5, 6)							
51. Gravel, loose and moist, 4 in. thick	0.25	0.60	0.65	0.70	0.75	0.80	0.70
52. Grass, marion bluegrass, 2 in. high	0.11	0.26	0.60	0.69	0.92	0.99	0.60
53. Snow, freshly fallen, 4 in. thick	0.45	0.75	0.90	0.95	0.95	0.95	0.90
54. Soil, rough	0.15	0.25	0.40	0.55	0.60	0.60	0.45
55. Trees, balsam firs, 20 sq ft ground area/tree, 8 ft high	0.03	0.06	0.11	0.17	0.27	0.31	0.15
56. Water surface, as in a pool	0.01	0.01	0.01	0.02	0.02	0.03	0.00

*See p. 32

‡Coefficients are per square foot of seating floor area or per unit. Where the audience is randomly spaced (e.g., in a courtroom, cafeteria, etc.), mid-frequency absorption can be estimated at about 5 sabins per person. Coefficients per person, however, must be stated in relation to spacing pattern to be precise.

§The audience area must be calculated to include an edge effect at aisles equal in area to a strip 3 ft wide for an aisle bounded on both sides by audience and a strip 1 1/2 ft wide for an aisle bounded on only one side by audience. No edge effect is used when the seating abuts walls or balcony fronts. The coefficients are also valid for orchestra and choral areas. Orchestra areas include people, instruments, music racks, etc., and no edge effects are used around musicians.

¶Coefficients for openings depend on absorption and volume of opposite side.

Test Reference

"Sound Absorption of Acoustical Materials in Reverberant Rooms," ASTM Method C423-66. American Society for Testing and Materials (ASTM), 1916 Race Street, Philadelphia, Pa. 19103.

Sources

1. Beranek, L. L.; Audience and Chair Absorption in Large Halls II, *J. Acoust. Soc. Amer.*, vol. 45, no. 1, January, 1969.
2. Evans, E. J., and E. N. Bazley; "Sound Absorbing Materials," H. M. Stationery Office, London, 1964.
3. Kingsbury, H. F., and W. J. Wallace; Acoustic Absorption Characteristics of People, *Sound Vib.*, vol. 2, no. 12, December, 1968.
4. Moore, J. E., and R. West; In Search of an Instant Audience, *J. Acoust. Soc. Amer.*, vol. 48, no. 6, December 1970.
5. "Performance Data, Architectural Acoustical Materials," Acoustical and Insulating Materials Association (AIMA), 205 West Touhy Avenue, Park Ridge, Ill. 60068. This bulletin is published annually.
6. Siekman, W.; Outdoor Acoustical Treatment: Grass and Trees, *J. Acoust. Soc. Amer.*, vol. 46, no. 4, October, 1969.

SOUND ABSORPTION: Suspended Sound-absorbing Units

Ref.: T. Mariner; Control of Noise by Sound-absorbent Materials, Noise Control, July, 1957, p. 15.

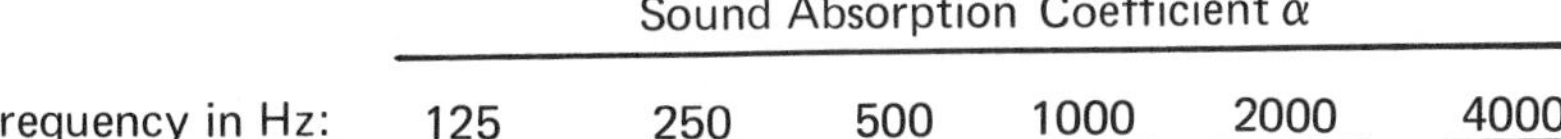

Sound Absorption Coefficient α

A. 1″-thick porous material

Frequency in Hz:	125	250	500	1000	2000	4000
α per sq ft of ceiling area	0.13	0.53	0.91	0.94	0.91	0.88

B. Suspended 1″-thick porous material

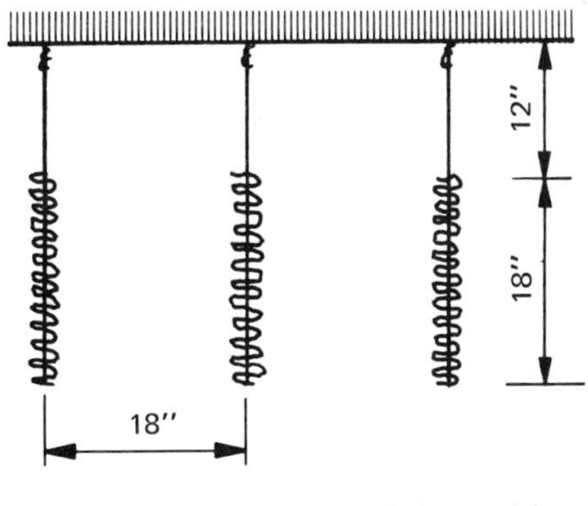

Frequency in Hz:	125	250	500	1000	2000	4000
α per sq ft of suspended material	0.07	0.20	0.40	0.52	0.60	0.67

C. Suspended 1″-thick porous material

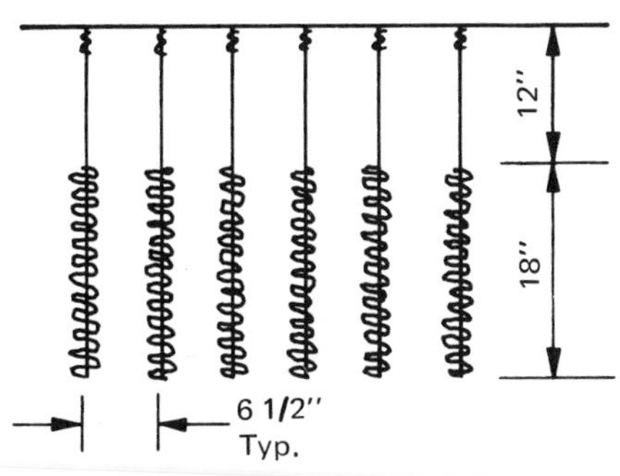

Frequency in Hz:	125	250	500	1000	2000	4000
α per sq ft of suspended material	0.10	0.29	0.62	1.12	1.33	1.38

NOTE: Total absorption in sabins from suspended units, as in B and C, is limited by the number of units that can be practically installed on a ceiling at the recommended spacings. See page 156 for a discussion of distributed absorbing material.

SOUND ABSORPTION: Commercially Available Sound-absorbing Materials

Some of the many commercially available sound-absorbing materials are shown below. Most sound-absorbing tiles and panels are not sufficiently durable for wall application. Consequently, protective and decorative "open" facings should be used. Some suitable examples are shown in this section. Be careful to observe the manufacturer's recommendations for the cleaning and painting of sound-absorbing materials. The AIMA booklet "How to Clean and Maintain Acoustical Tile Ceilings" presents useful guidelines.

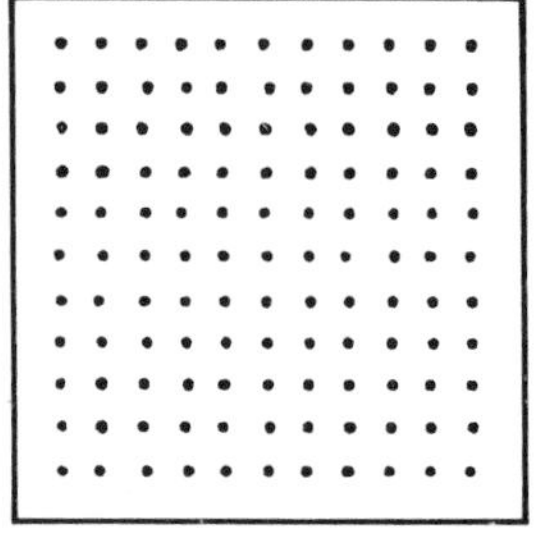

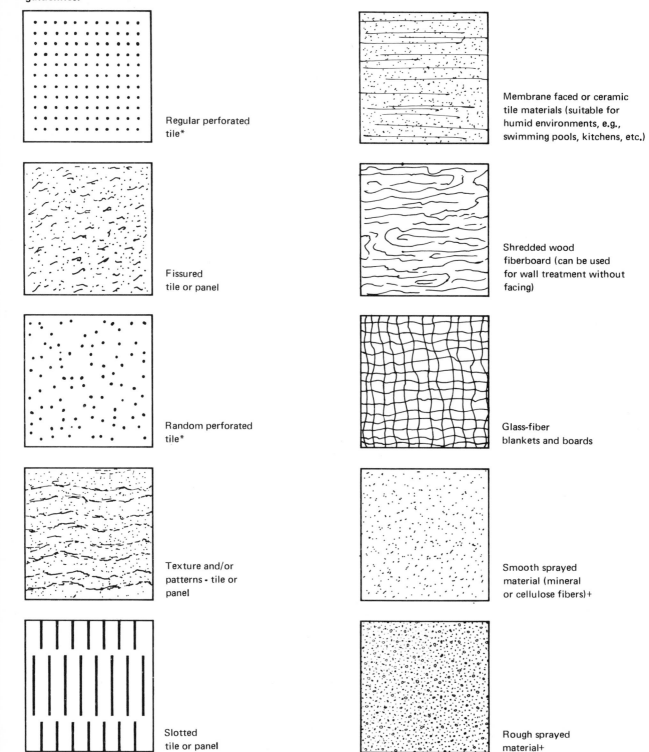

Regular perforated tile*

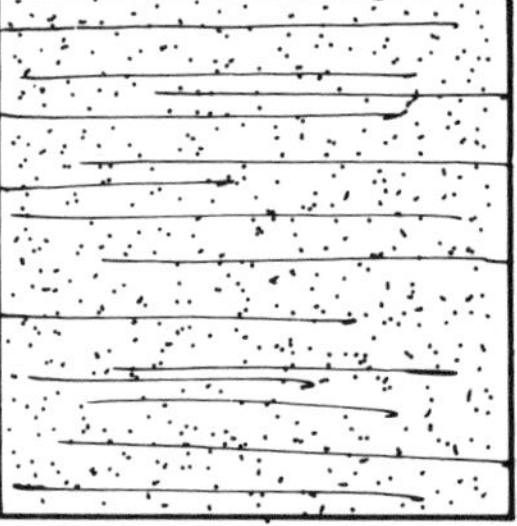

Membrane faced or ceramic tile materials (suitable for humid environments, e.g., swimming pools, kitchens, etc.)

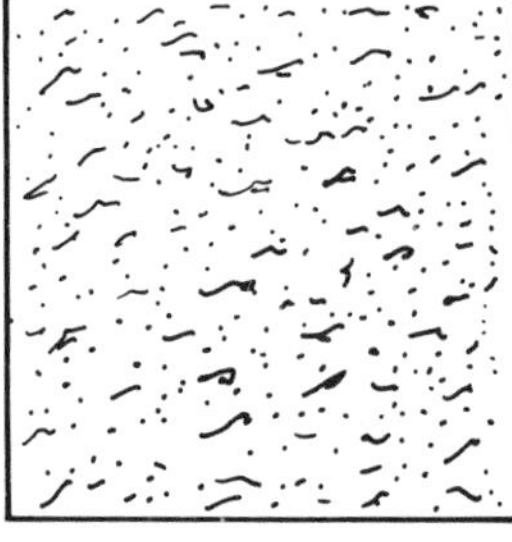

Fissured tile or panel

Shredded wood fiberboard (can be used for wall treatment without facing)

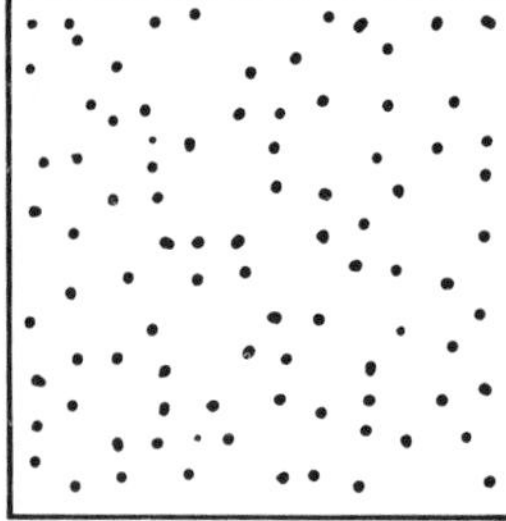

Random perforated tile*

Glass-fiber blankets and boards

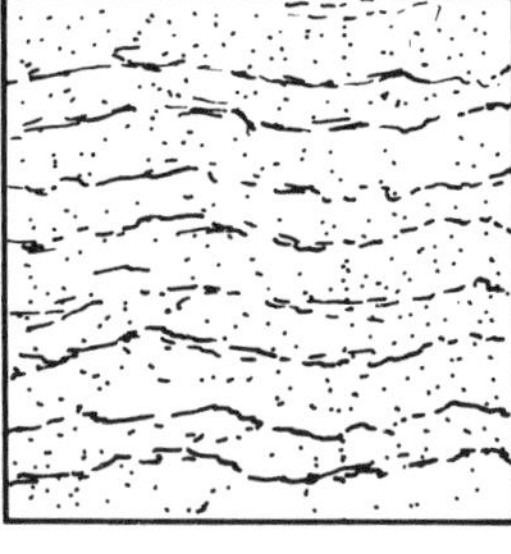

Texture and/or patterns - tile or panel

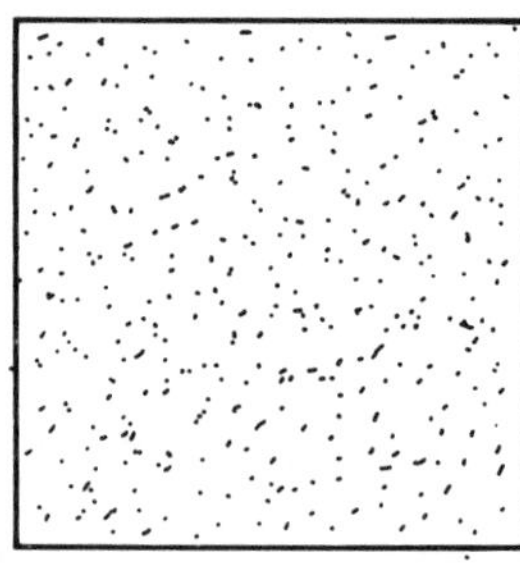

Smooth sprayed material (mineral or cellulose fibers)+

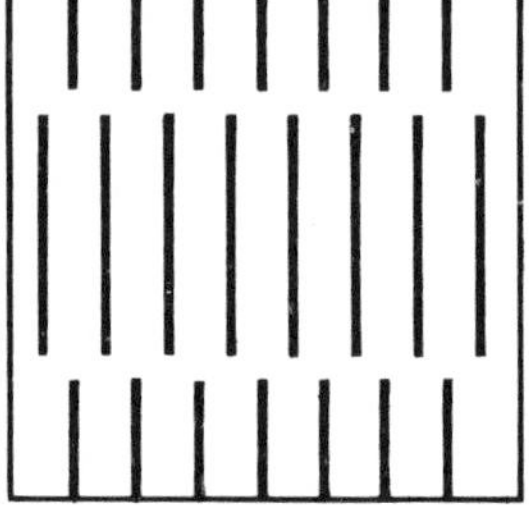

Slotted tile or panel

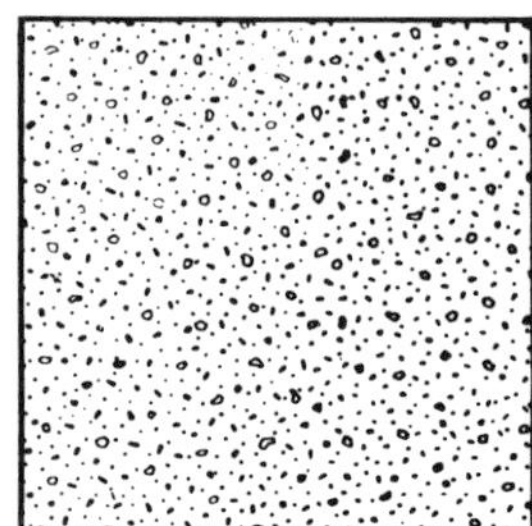

Rough sprayed material+

* Openings provide about 15% open area to allow painting without bridging over the holes—avoid oil and rubber-base paints which may clog pores.

† Use sprayed materials in 1" to 3" thicknesses on hard backup surface or apply to open lath.

SOUND ABSORPTION: Laboratory Test Mountings for Sound-absorbing Materials

Sound absorption laboratory tests should be conducted according to current ASTM procedures. The types of mountings shown below are typical of actual installation methods used in the field. Mtgs. Nos. 1, 2, 4, and 7 apply to most prefabricated products, No. 6 to sound-absorbing mechanical duct linings, and Nos. 5 and 8 are used for specialized applications. Be sure the mounting used is always given with the product's sound absorption coefficients!

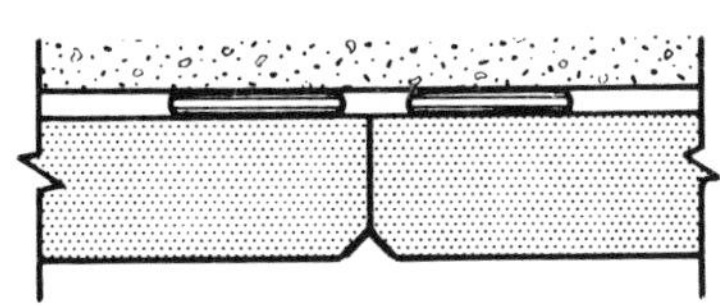

1. Cemented to plasterboard with 1/8" air space. Considered equivalent to cementing to plaster or concrete ceiling.

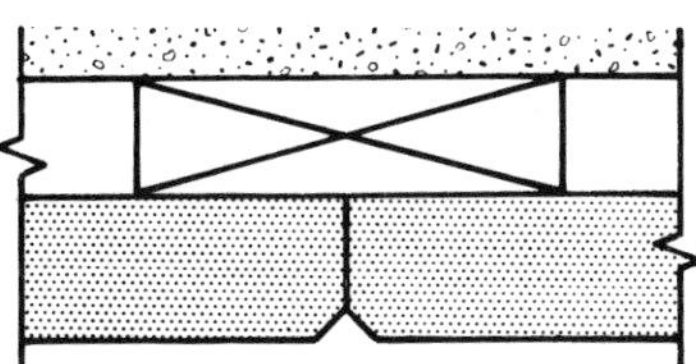

2. Nailed to nominal 1" x 3" (3/4" x 2 5/8" actual) wood furring 12" o.c.

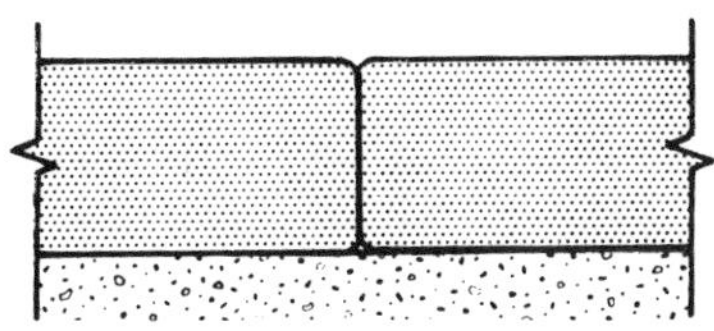

4. Laid directly on laboratory floor.

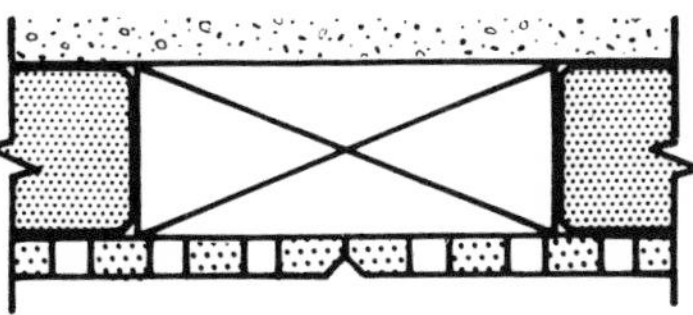

5. Wood furring 1" x 3" (3/4" x 2 5/8" actual) 24" o.c. 1" mineral wool between furring unless otherwise indicated. Perforated facing fastened to furring.

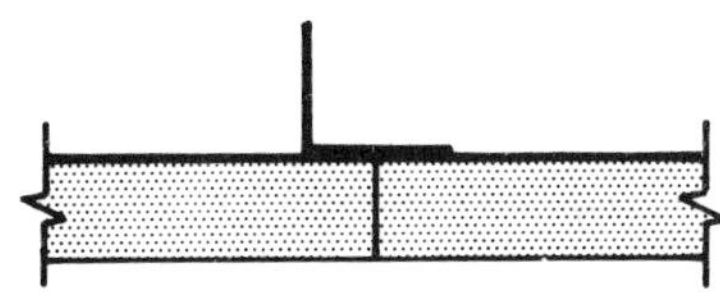

6. Attached to 24-ga. sheet iron, supported by 1" x 1" x 1/8" angle iron.

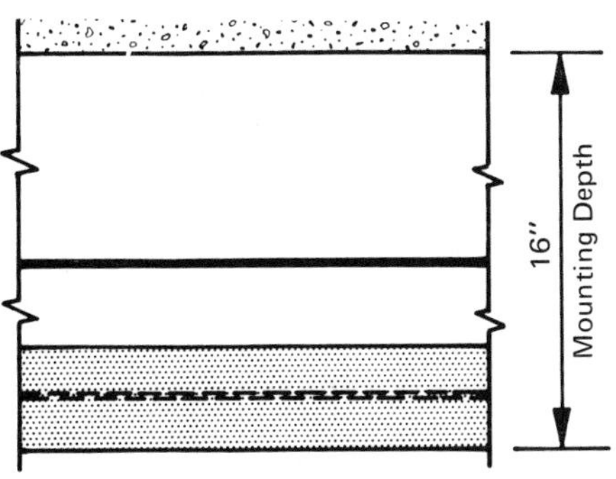

7. Mechanically mounted on special metal supports.

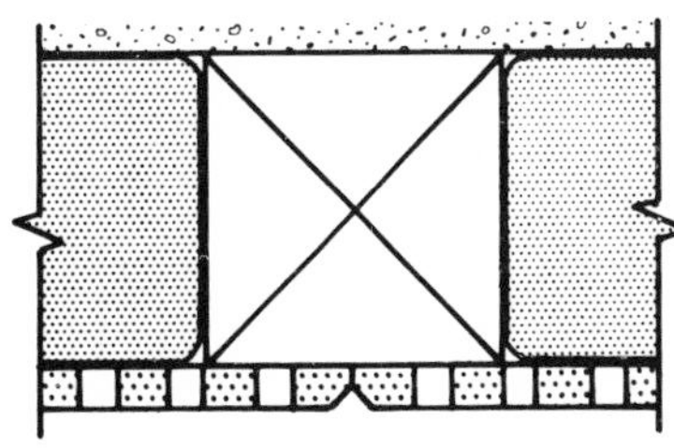

8. Wood furring 2" x 2" (1 5/8" x 1 5/8" actual) 24" o.c. 2" mineral wool between furring. Perforated facing fastened to furring.

X-00. A spaced mounting similar to No. 7 except that mounting depth is variable.

Courtesy of Acoustical and Insulating Materials Association.

USES OF SOUND-ABSORBING MATERIALS

Reverberation Control

Sound-absorbing materials are used to *control reverberation*. Each doubling of the total amount of absorption in a room reduces the reverberation time by one-half. In small rooms sound-absorbing materials reduce the multiple reflections of sound from the room surfaces and also make the sound seem to come directly from the actual source rather than from everywhere in the room.

Noise Reduction in Rooms

When correctly used, sound-absorbing materials can also be effective in *controlling noise* within a room. They have a limited application here and are not, as too many would believe, the panacea for all noise problems. For example, each doubling of the total amount of absorption in a room results in the noise being reduced by only 3 dB. One can see that, as with other aspects of sound, the law of diminishing returns can quickly limit the effectiveness of this solution.

Echo Control

Additionally, sound-absorbing materials are used to *control echoes* (usually simultaneously with controlling reverberation). Echoes are long-delayed discrete sound reflections of sufficient intensity to be clearly heard above the general reverberation in a room. A disturbing phenomenon called "flutter echo" may be present in small rooms and also can be controlled with sound-absorbing materials.

REVERBERATION TIME

Until the pioneering work of Wallace Clement Sabine, starting about 1895, criteria for good listening conditions in rooms were largely nonexistent. Professor Sabine had been asked to attempt to improve the atrocious listening conditions of the lecture hall in Harvard University's new Fogg Art Museum. Sound in the hall would persist for about 5 1/2 sec due to multiple reflections from the hard-surfaced room finish materials. In 5 1/2 sec most persons can utter about 15 syllables. Consequently, speech intelligibility was almost impossible everywhere in the hall.

Sabine recognized that the problem of the persistence of reflected sound was due to the size and the furnishings of a room. He conducted tests in the hall, using an organ pipe with an initial sound intensity level of 60 dB above the hall's ambient noise level at a frequency of 512 Hz (cps in those days). Sabine's disciplined sense of hearing was used to judge when the sound ceased to be audible. The time it took sound to decay 60 dB was measured by a stopwatch and defined by Sabine as the reverberation time. Sabine was able to test only at night when it was relatively quiet – after the streetcars stopped running around midnight and before the milkmen started rattling their carts over the cobblestones in the early morning hours. By bringing in seat cushions (with student help) from nearby Sanders Theatre, he found that he could lower the reverberation time to about 1 sec. These cushions were made of porous sound-absorbing vegetable fiber material which was covered with a canvas cloth. The greater the number of cushions brought in, the greater the total room absorption and therefore the reduction of the reverberation. Consequently, the first unit of sound absorption was the absorption of one seat cushion of the Sanders Theatre!

The results of his work made it possible for the first time to program optimum reverberation time in advance of construction. Today, desired room acoustics can be the result of design, not chance or faithful reproduction. The expression for reverberation time, which Sabine defined and proved empirically, is

$$T = 0.05 \frac{V}{a}$$

where T = reverberation time, sec (or time required for sound to decay 60 dB after source has stopped)
V = room volume, cu ft
a = total room absorption, sabins (i.e., square feet of absorption)

Since the above formula (often referred to as the Sabine equation) is generally used by the measurements laboratories today to compute absorption coefficients, its use is appropriate for most architectural work. The preferred range of reverberation times, based on experience with completed spaces, is shown on the following page as a black bar for the given function. The range is also extended with the dashed sections, which indicate the extremes of acceptability.

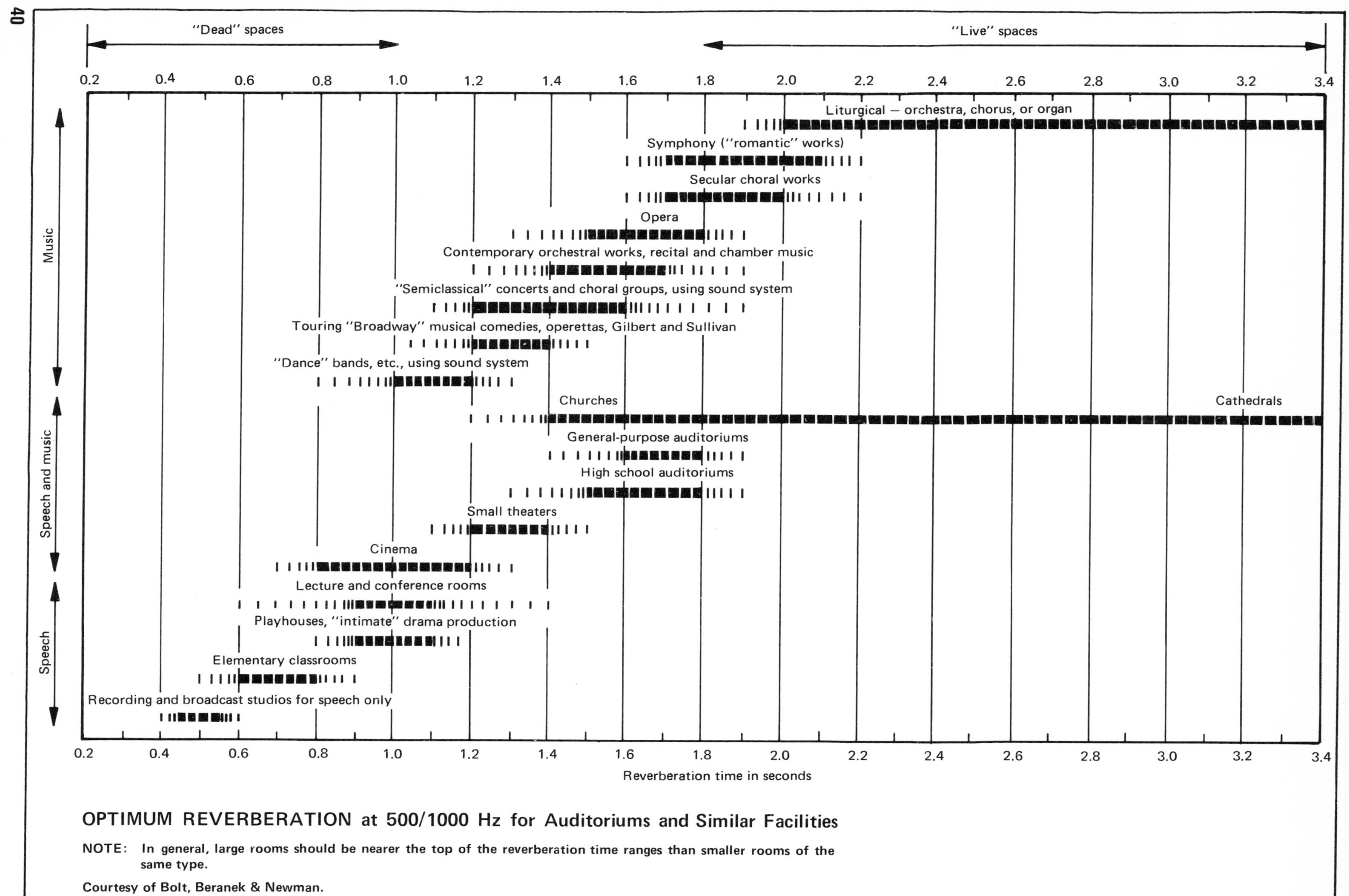

OPTIMUM REVERBERATION at 500/1000 Hz for Auditoriums and Similar Facilities

NOTE: In general, large rooms should be nearer the top of the reverberation time ranges than smaller rooms of the same type.

Courtesy of Bolt, Beranek & Newman.

SOUND ABSORPTION: Example Problem—Reverberation Time

Given:
School classroom 60 x 35 x 15 ft high
Sound absorption coefficients at 500 Hz are:

α walls = 0.30
α ceiling = 0.05
α floor = 0.10

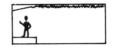

Section

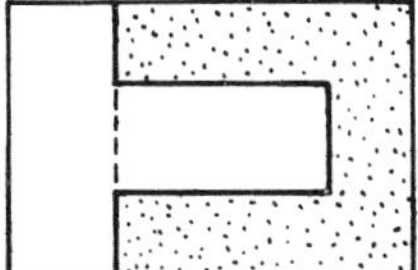
Reflected ceil. plan

(1.) What is the reverberation time in this space with no acoustical treatment?

(2.) What will the reverberation time be if 50% of the ceiling surface, along the perimeter, is treated with acoustical tile (α at 500 Hz = 0.85)?

(1.) Room volume V = 60 X 35 X 15 = 31,500 ft^3, find absorption (a) in sabins:

$$T = 0.05\ V/a = \frac{0.05 \times 31{,}500}{a} = \frac{1575}{a}$$

	Area, ft^2	α	a, sabins
Ceiling	60 X 35 = 2100	0.05	105
Walls	2 X 35 X 15 = 1050		
	2 X 60 X 15 = 1800		
	2850	0.30	855
Floor	60 X 35 = 2100	0.10	210
		Total a =	1170 sabins

T = 0.05 V/a = 1575/1170 = **1.35 sec** at 500 Hz (unoccupied)

(2.)

	Area, ft^2	α	a, sabins
Bare ceiling	1050	0.05	53
Treated ceiling	1050	0.85	892
Walls	2850	0.30	855
Floor	2100	0.10	210
		Total a =	2010 sabins

T = 0.05 V/a = 1575/2010 = **0.78 sec** at 500 Hz (unoccupied)

ROOM NOISE REDUCTION CHART

The buildup of sound levels in a room is due to the repeated reflections of sound from the enclosing surfaces. This buildup is affected by the amount of absorption within the room. The difference in decibels in reverberant field noise levels (SPLs or ILs) under two conditions of room absorption can be found as follows:

$$NR = 10 \log \frac{a_2}{a_1}$$

where NR = room noise reduction, dB
a_2 = total room absorption after treatment, sabins
a_1 = total room absorption before treatment, sabins

The chart below also can be used to determine the reduction of reverberant noise level within a room due to changing the total room absorption. For example, if the total amount of absorption in a space can be increased from 700 to 2100 sabins, the reduction in reverberant noise level (NR) will be about 5 dB (see dot on chart scale at absorption ratio $a_2/a_1 = 3$). Since absorption coefficients vary with frequency, the NR should be calculated at the several frequencies for which coefficients are given.

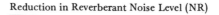

Reduction in Reverberant Noise Level (NR)

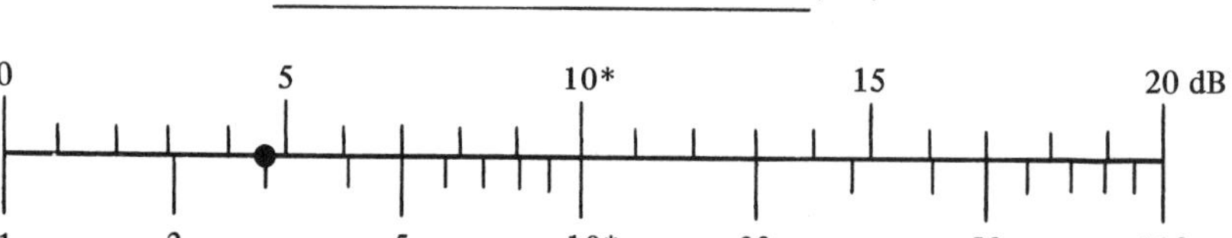

Ratio of Total Room Absorption (a_2/a_1)

*Practical upper limit of improvement

Note that the NR is the reduction in reverberant noise level and does not affect the noise level very near the source of sound in a room. Also, as indicated on the chart, a reduction in reverberant noise level of 10 dB (an increase in absorption of greater than 10 times the initial value before treatment) is the practical upper limit for most remedial situations.

SOUND ABSORPTION: Example Problem—Room Noise Reduction

Assume a room 10′ × 10′ × 10′ finished in concrete. The absorption coefficient of concrete is $\alpha = 0.02$ in the middle frequencies. Find the noise reduction in the room at 500 Hz only in this example; but in most design situations, one should also make computations at other important frequencies. The total room surface area S is 600 sq ft. The absorption coefficient assumed for acoustical tile at 500 Hz is $\alpha = 0.70$.

First, consider the room with no acoustical treatment.

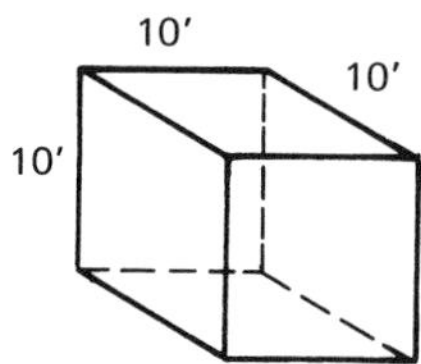

S = 600 sq ft and α for concrete = 0.02

$a_1 = \Sigma S \alpha$

$a_1 = 600\ (0.02) =$ **12 sabins**

Now, cover the ceiling with acoustical tile.

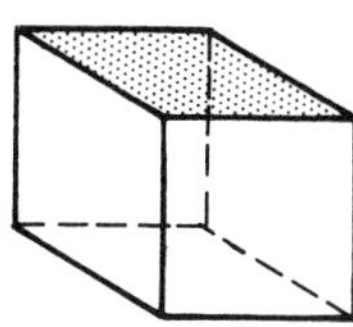

$a_2 = \Sigma S \alpha$

$a_2 = 500\ (0.02) + 100\ (0.70) = 80$ sabins

$NR = 10 \log a_2/a_1$

$= 10 \log 80/12$

$= 10 \log (6.66)$

$= 10\ (0.82)$

NR = 8 dB

If this were a noisy shop, the reverberant noise would be 8 dB less after adding the acoustical tile, which is a significant improvement.

Now, also cover the four walls with acoustical tile.

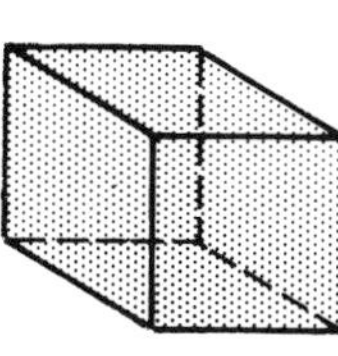

$a_3 = \Sigma S \alpha$

$a_3 = 100\ (0.02) + 500\ (0.70) = 352$ sabins

$NR = 10 \log a_3/a_1$

$= 10 \log 352/12$

$= 10 \log (29.3)$

$= 10\ (1.47)$

NR = 15 dB

SUMMARY:

Surfaces Covered with Fuzz	Room Noise Reduction at 500 Hz
1. Ceiling only	8 dB
2. Ceiling + one wall	11 dB
3. Ceiling + four walls	15 dB, which is about the maximum NR possible by adding fuzz to a room and would not be as great at other frequencies.

SOUND ABSORPTION: Relative Effectiveness of Wall and Ceiling Absorption Treatment

Noise control and reverberation control treatments can be placed on any available surface; the important goal is to provide sufficient absorption. The examples below show that high-efficiency absorptive treatment of walls can be more effective in smaller rooms, whereas treatment of ceilings is more effective in larger rooms.

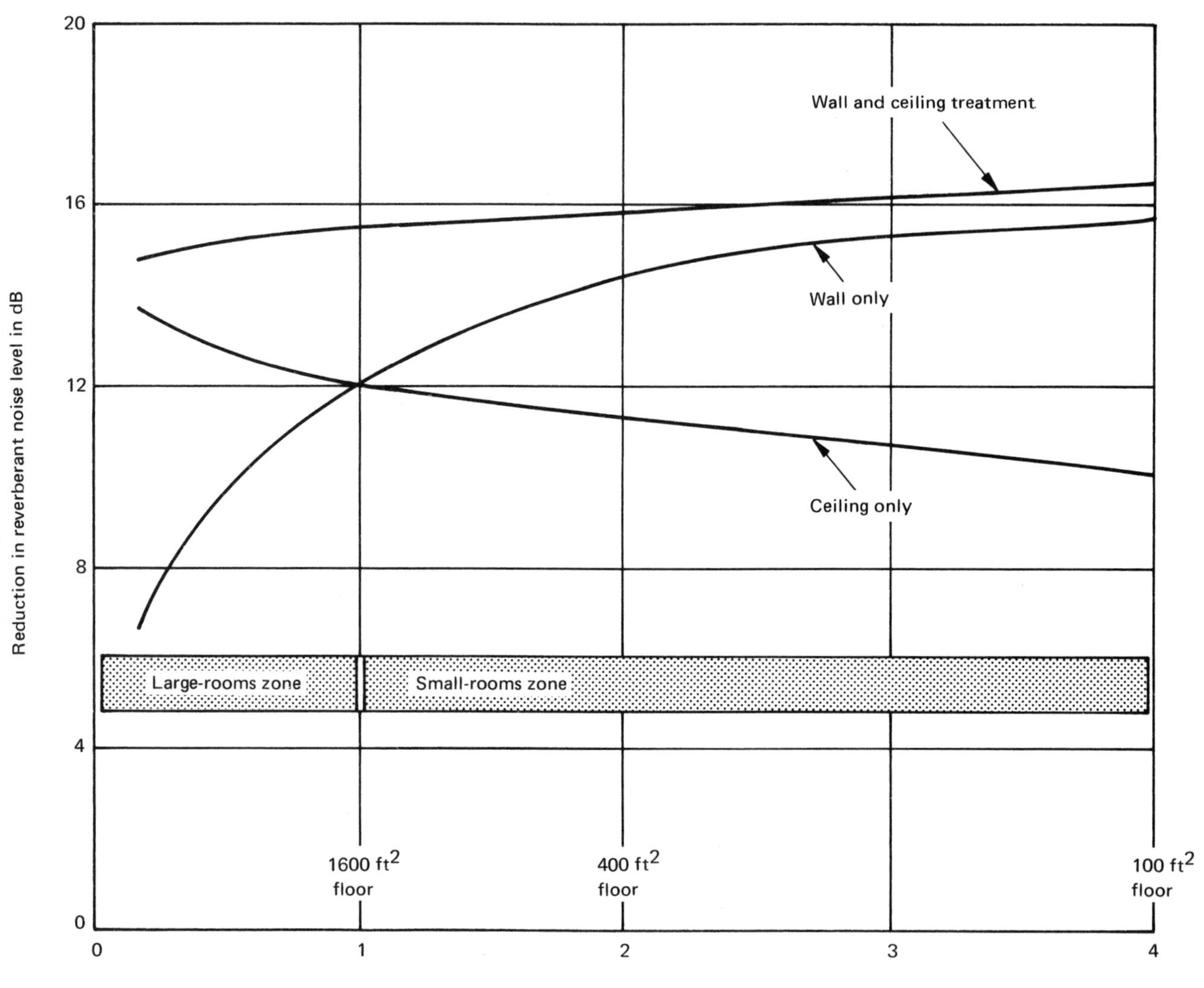

NOTE: Absorption coefficient of untreated surface is assumed equal to 0.02; ceiling height assumed at 10′-0″.

SOUND ABSORPTION: Open Facings to Protect "Fuzz" Wall Treatment

Surface area of facing should be at least 20% open for reverberation or room noise control, where high frequency absorption may not be critical. So many possibilities will satisfy these requirements that wall treatment is often limited only by the designer's imagination.

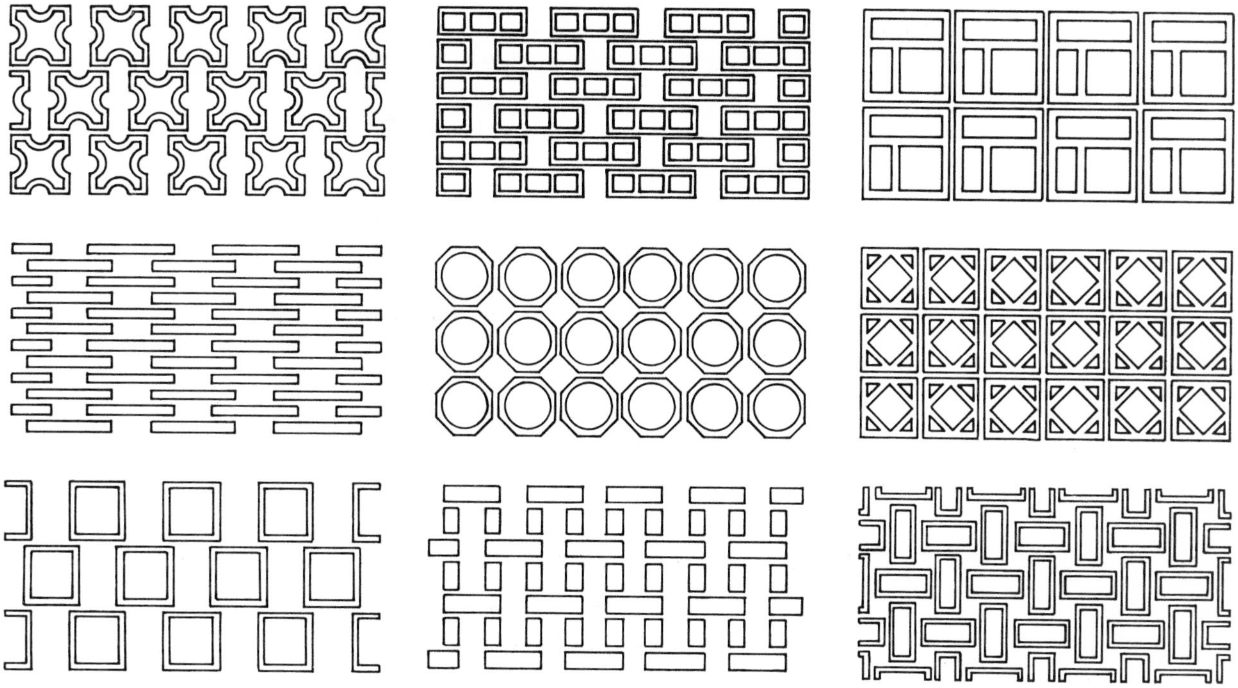

NOTE: See page 32 for typical sound absorption data from porous material with open facing.

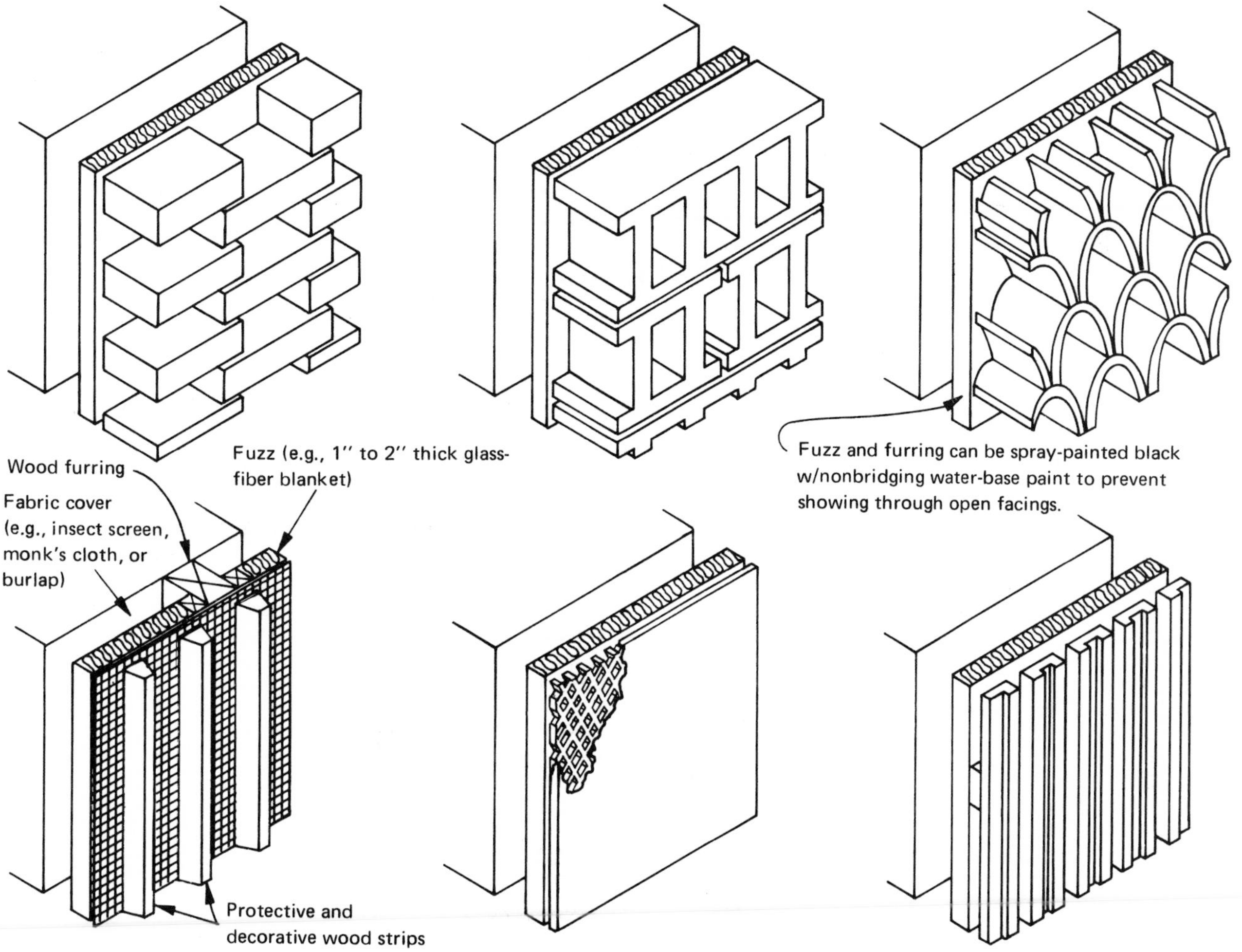

Wood furring

Fuzz (e.g., 1" to 2" thick glass-fiber blanket)

Fabric cover (e.g., insect screen, monk's cloth, or burlap)

Protective and decorative wood strips

Fuzz and furring can be spray-painted black w/nonbridging water-base paint to prevent showing through open facings.

Ref.: R. B. Newman, and W. J. Cavanaugh, Acoustics in J. H. Callender (ed.), Time-saver Standards, 4th ed., Mc Graw-Hill, New York, 1966.

SOUND ABSORPTION: Acoustically Transparent Facings

Acoustically transparent facings may range from 5 to 50% or more open area, depending on the absorption requirements. As a rule, facings tend to reduce or "cut off" high frequency sound-absorbing effectiveness, subject to the % open, dimensions of solid area, etc.

Perforated materials such as perforated sheet metal, expanded metal, or punched and pressed metal can be used alone in front of "fuzz," or together with wood slats or other large-scale elements, as shown on the preceding page. Typical open metal materials are shown below along with a table of suitable perforation sizes and spacings for general facing materials.

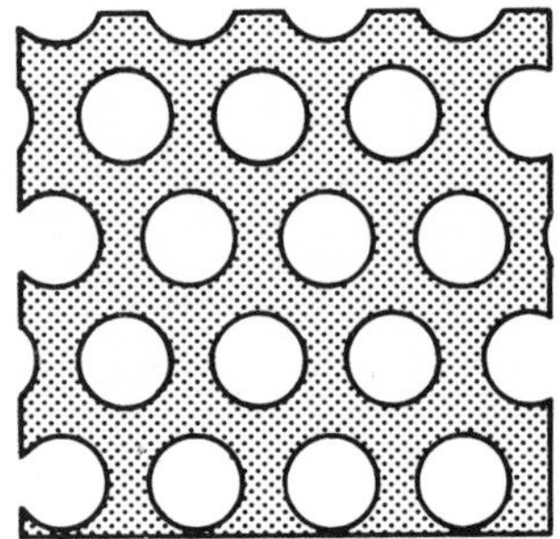

1/4" Staggered holes
at 3/8" o.c. - 40% open

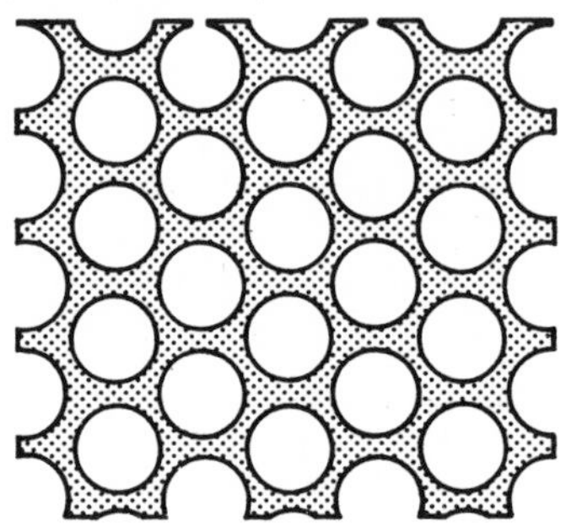

1/4" Staggered holes
at 5/16" o.c. - 58% open

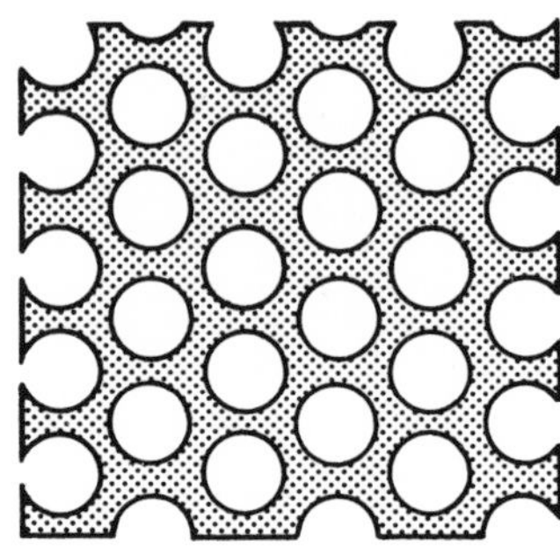

17/64" Staggered holes
at 5/16" o.c. - 65% open

Suitable perforation sizes and spacings

Hole diameter (inch)	Spacing (inch o.c.)	Notes
3/16	0.50	Safe limit for hardboard ("pegboard") material
5/32	0.40	
1/8	0.30	Most suitable for wall materials. Can be painted without clogging holes, and holes are small enough to discourage the jabbing of sharp objects into them.
3/32	0.22	
1/16	0.15	
1/32	0.08	

SOUND ABSORPTION: Sound-absorbing Ceiling Arrangements

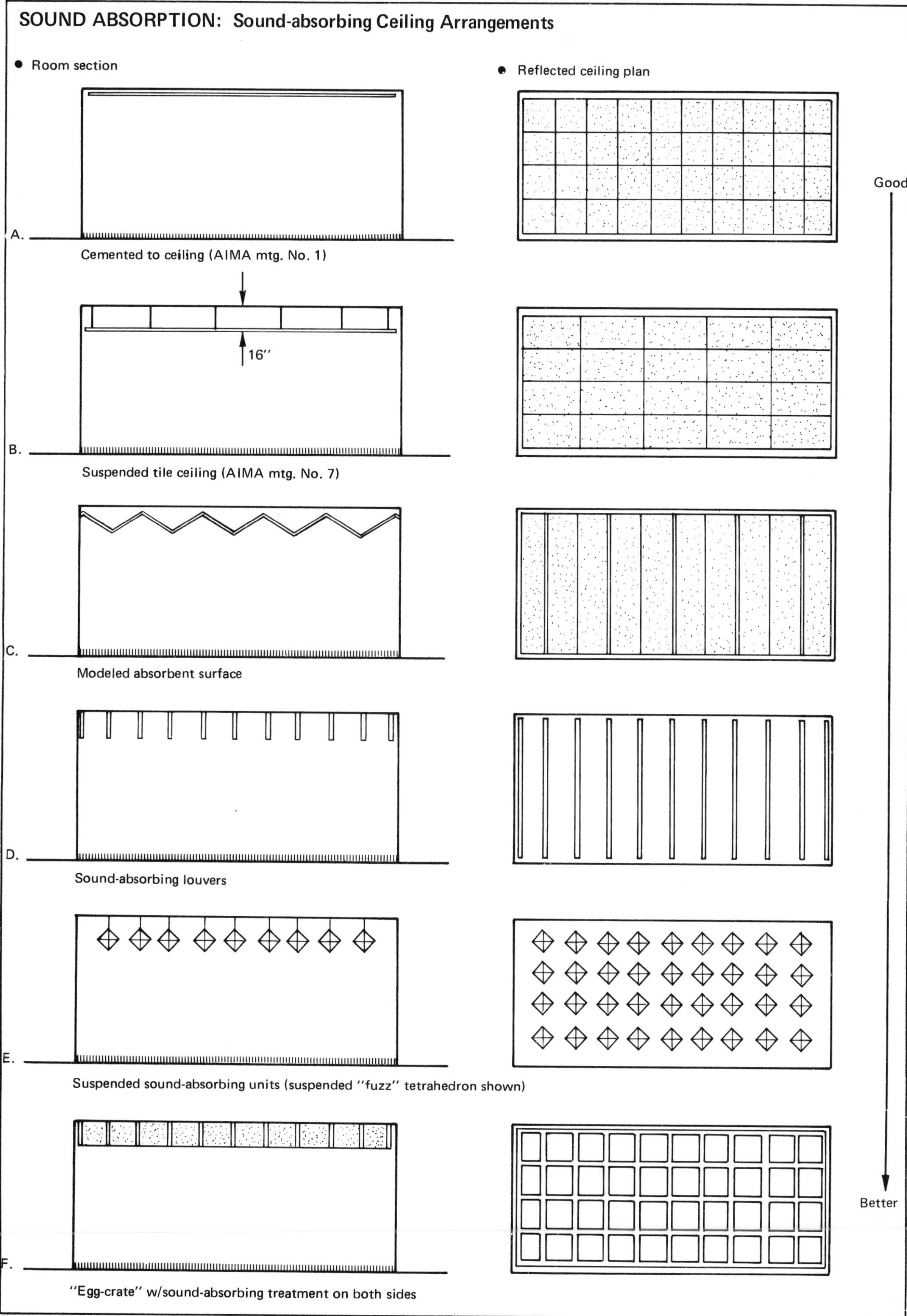

INDUSTRIAL NOISE EXPOSURE

In 1969 the U.S. Department of Labor, through the Walsh-Healey Public Contracts Act (*Federal Register*, May 20, 1969), adopted standards for occupational noise exposure. These rules and regulations apply to any manufacturer who sells goods (or services) to the Federal government having a value in excess of certain nominal amounts. The permissible noise exposures are as follows:

Duration per Day, hr	*Sound Pressure Level, dBA*
8	90
6	92
4	95
3	97
2	100
1 1/2	102
1	105
1/2	110
1/4 or less	115

In addition, exposure to impulsive noise such as gunfire, or impact noise from machinery, etc., should not exceed 140 dBA peak sound pressure level.

The sound pressure level limitations above are in terms of dBA, which refers to the A-scale reading on a sound-level meter. Note that amplified rock-and-roll music at 120 dBA would exceed the shortest permissible noise exposure given in the above table! Sound-level meters are usually equipped with weighting networks that tend to represent the frequency characteristics of the average human ear for various sound levels. Even though this government regulation is presented in terms of dBA exposures, noise control measures should normally be designed in terms of octave-band data by following the procedures outlined throughout this book. Octave-band analysis of a noise gives a picture of how severe a problem is at different frequencies, which in turn helps determine the appropriate corrective measures to be taken. For example, remedial solutions for low frequency noise problems differ considerably from those for high frequency.

SOUND ABSORPTION: Room Noise Reduction

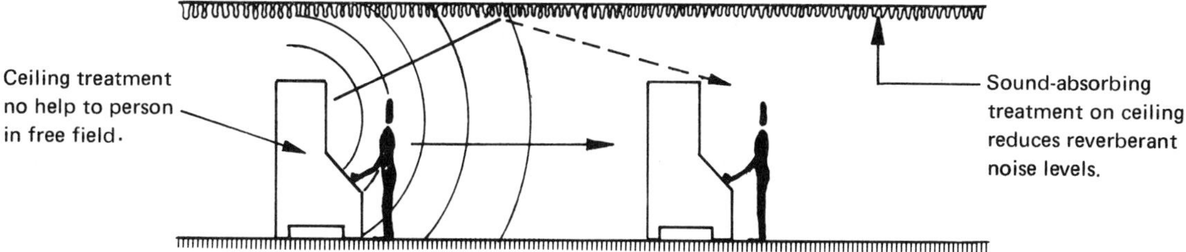
Ceiling treatment no help to person in free field.
Sound-absorbing treatment on ceiling reduces reverberant noise levels.

● Low ceiling, machines widely spaced

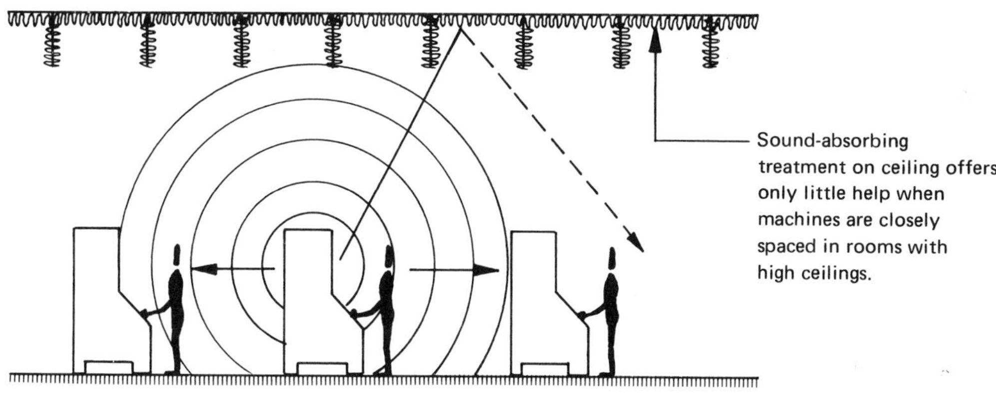
Sound-absorbing treatment on ceiling offers only little help when machines are closely spaced in rooms with high ceilings.

● High ceiling, machines closely spaced

NOTE: **Even though sound-absorbing treatment on ceiling does not reduce noise levels in the free field, it is important because (1) the reverberant noise levels may be higher than the free field noise of some machines, and (2) a reduction in reverberation time will help make machine noise more directional, allowing workers to be more attentive to their own machines.**

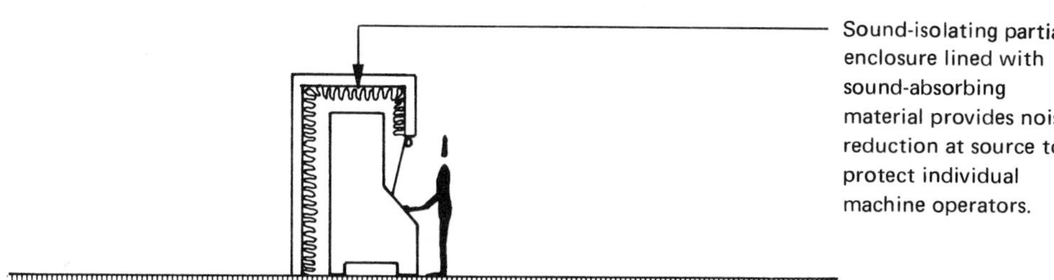
Sound-isolating partial enclosure lined with sound-absorbing material provides noise reduction at source to protect individual machine operators.

● Enclosure to contain machine noise

Ref.: R. B. Newman, and W. J. Cavanaugh, Design for Hearing, Progr. Architect., May, 1959.

USE OF SOUND POWER LEVEL DATA

Sound power level (PWL) expresses the amount of sound or noise power that is radiated by a given source, regardless of the space into which the source is placed. The noise level (SPL) in that space can be estimated by knowing the space's acoustical characteristics from the total room absorption and volume.

Example: A 30-hp electric motor is to be placed in an unfurnished 10-by 10-by 10-ft painted concrete mechanical equipment room. The motor manufacturer furnished the PWL data for his equipment listed below.

OCTAVE-BAND CENTER FREQUENCIES	125 Hz	250 Hz	500 Hz	1000 Hz	2000 Hz	4000 Hz
PWL (dB reference 10^{-12} watt*)	88	92	93	93	92	86

To find the noise level from the motor in the bare concrete room, the room absorption is calculated and the effects of PWL estimated from the chart on the following page.

	125 Hz	250 Hz	500 Hz	1000 Hz	2000 Hz	4000 Hz
Room absorption, sabins	60	30	36	42	54	64
Reverberation time, sec	0.8	1.7	1.4	1.2	0.9	0.8
SPL-PWL (dB) from chart on following page	–1	2	1	0	–1	–1
Reverberant SPL, dB	87	94	94	93	91	85

If the walls and ceiling are treated with 2 in. fuzz, the reduced noise level (SPL) can be found as follows:

	125 Hz	250 Hz	500 Hz	1000 Hz	2000 Hz	4000 Hz
New room absorption, sabins:						
Floor	10	5	6	7	9	8
Walls and ceiling	190	300	390	400	390	350
Total sabins	200	305	396	407	399	358
Reverberation time, sec	0.25	0.20	0.12	0.12	0.12	0.14
SPL-PWL (dB) from chart	–6	–8	–9	–9	–9	–9
Reverberant SPL, dB	82	84	84	84	83	77

*It is extremely important to know the reference power since many manufacturers, until recent years, presented data with 10^{-13} watt as the reference value. One can see that a 10 dB error could easily result if the manufacturer's reference were unknown.

SOUND ABSORPTION: Reverberant Sound Pressure Levels from Sound Power Levels

EXAMPLE – Use of Graph:

Given: 50 hp pump with a PWL of 90 dB (re: 10^{-12} watt) at 500 Hz. Room volume of 5,000 ft^3 and reverberation time (T) of 1 sec at 500 Hz.

Procedure to find SPL:
1. Enter graph at volume of 5,000 ft^3 and read opposite T = 1.0 sec curve to $\simeq -7$ dB (see dashed lines on graph).
2. Therefore, SPL − PWL = −7 and SPL = 90 − 7 = 83 dB at 500 Hz.

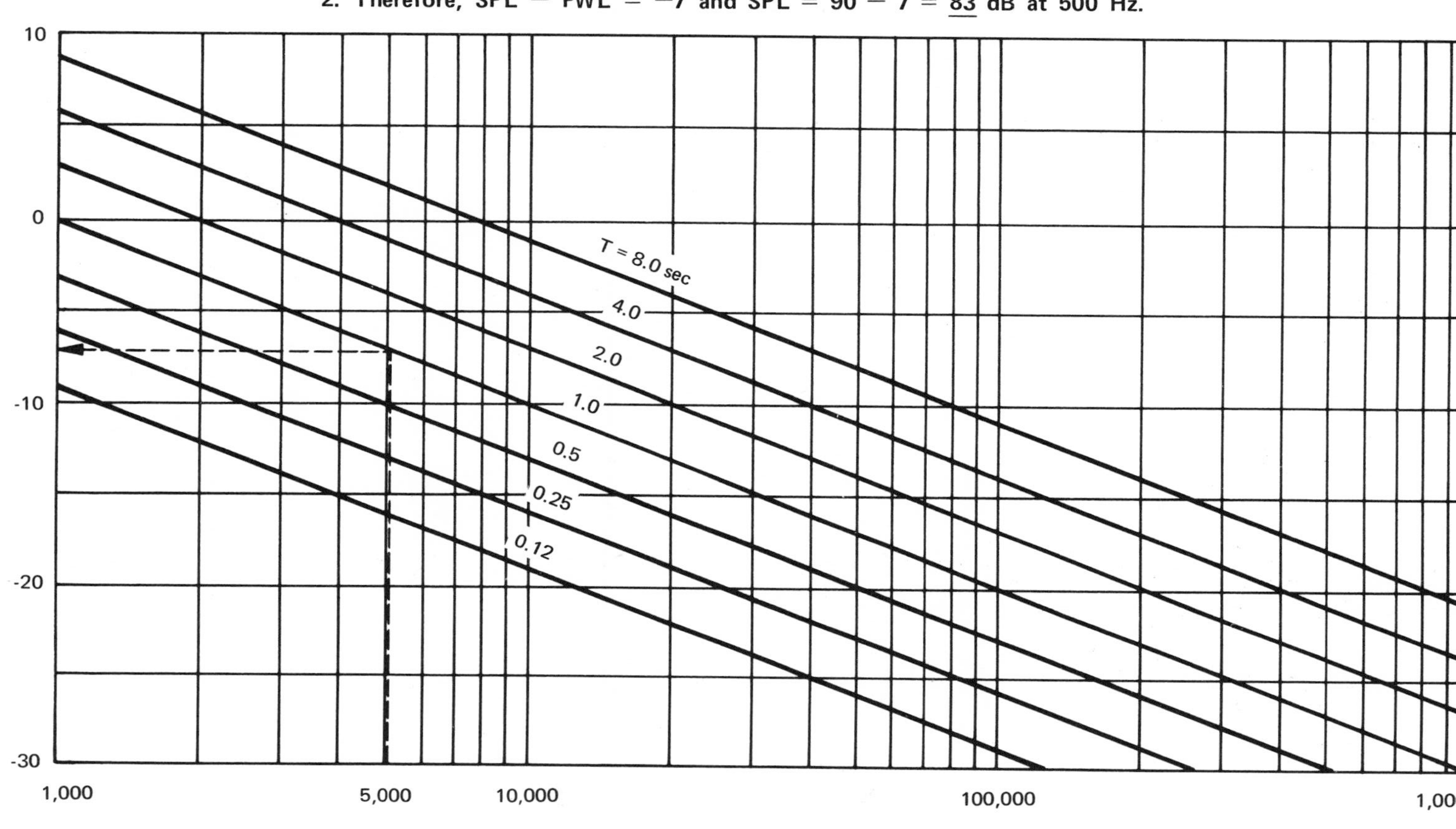

SUGGESTED SOUND-ABSORBING TREATMENT FOR ROOMS

Although the NRC number has the limitations presented earlier in this section, it can be an adequate index for the treatment of the noncritical spaces listed below. The last group, however, is a sample of the many spaces where the NRC by itself is inadequate and special study is required.

Type of Space	Preferred NRC Range	Ceiling Treatment	Wall Treatment
Bedrooms, sleeping quarters, etc.	0.45–0.65	Full	None required
Private offices, large offices, small conference rooms, hospitals, laboratory work spaces, libraries, retail shops and stores, etc.	0.65–0.75	Full	None required
Lobbies, corridors, gymnasiums, etc.	0.65–0.75	Full	Yes
Secondary and college classrooms, large meeting rooms, etc.	0.65–0.75	Partial	Yes
Kitchens, cafeterias, laundries, restaurants, etc.	Greater than 0.75	Full	Usually none required
Computer equipment rooms, school and industrial shops, machinery spaces, etc.	Greater than 0.75	Full	Yes
Auditoriums, theaters, radio/TV studios, music practice rooms, audio-visual facilities, churches, courtrooms, chapels, mechanical equipment rooms, open plan offices and schools, language laboratories, factories, etc.	(These spaces in particular require special study to determine the appropriate type, amount, and location of sound-absorbing treatment.)		

CHECKLIST FOR EFFECTIVE ABSORPTION OF SOUND

- Never put sound-absorbing material on a surface that is needed for useful sound reflections.
- Place sound-absorbing material on any surface that can be expected to produce annoying echoes or to focus sound.
- In general, cover ceilings for noise reduction within rooms, unless the floor is carpeted and the room is filled with draperies and heavily upholstered furniture. Sound-absorbing materials are commercially available that have a factory-applied surface finish which is reasonably durable for ceiling applications besides satisfying appearance, light reflectance, and other architectural considerations.
- In long, narrow, or very high rooms, consider using absorption on the walls. In very large rooms with low ceilings, wall absorption is rarely beneficial.
- Generally, the construction of the building determines the mounting method; check carefully so the mounting used is best suited for the absorption desired. The actual method of mounting is important since it will control absorption efficiency. For example, sound-absorbing materials applied with adhesive (mounting No. 1) are poor low frequency abosrbers. However, when applied to furring systems (mounting No. 2), they will give somewhat better low frequency absorption; and when used in suspended ceiling systems (mounting No. 7), they will provide balanced absorption.
- The amount of treatment is determined by the absorbing material already in the room, plus the size of the room.

SECTION 3

Sound Isolation

SOUND ISOLATION: Doorbell Enclosure Demonstration

An inexpensive doorbell can be used to demonstrate the basic principles of sound isolation from the four conditions shown below. Good isolation is provided by massive and impervious materials!

Condition:

1. **No isolation**
2. **3/4" porous "fuzz" (Sound absorbents are very poor isolators, as air molecules can pass right through them.)**
3. **1/2" plywood w/airtight seal (Plywood is effective because it has mass, and the seal is very important as the smallest opening increases the transmitted sound considerably.)**
4. **3/4" porous "fuzz" plus 1/2" plywood w/airtight seal (Fuzz in the enclosure helps reduce reverberant sound near the source.)**

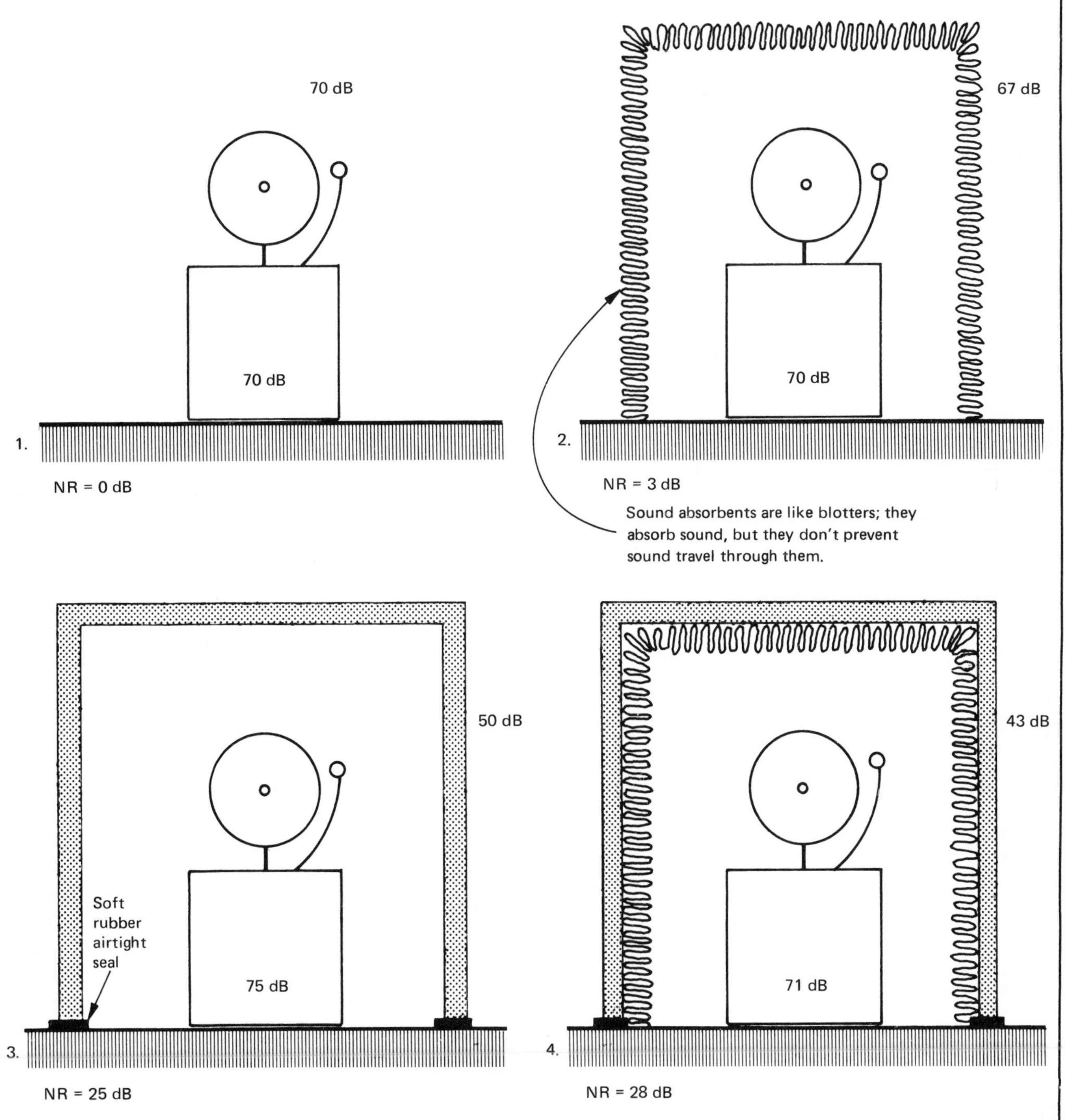

SOUND ISOLATION: Harbor Breakwater Analogy

Reductions in ocean wave energy from a concrete sea wall and a wood pile group are shown below. This is similar to the sound transmission process where transmission loss (TL) is the amount in decibels by which sound is reduced by passing from one side of a structure to the other. The more massive and solid the structure, the greater is its resistance to motion (according to Newton's II law, F = mA) and consequently the higher the TL.*

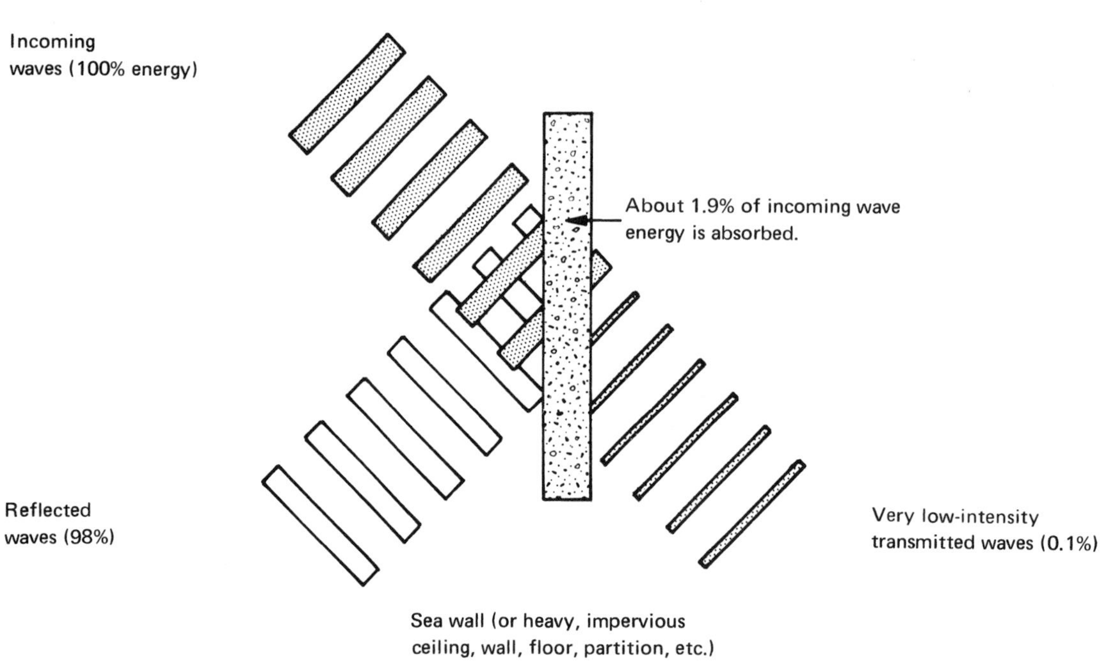

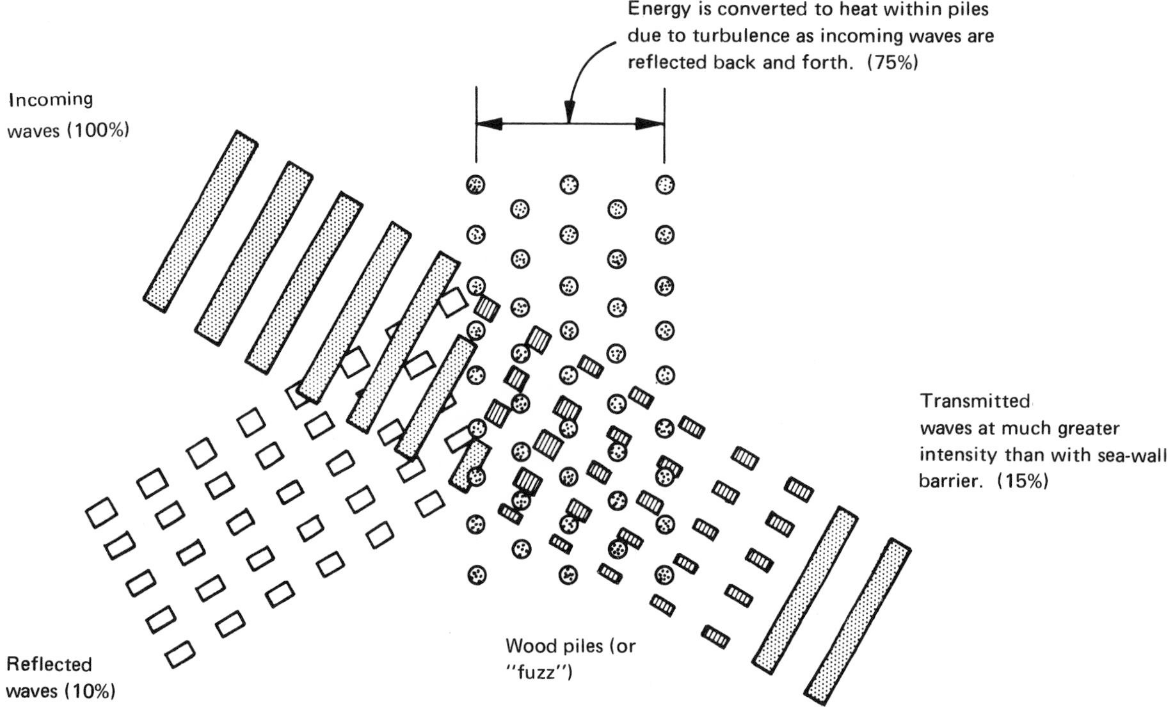

***TL can be expressed as 10 log 1/τ, where τ is the sound transmission coefficient from laboratory tests. As with other aspects of sound, TL varies with frequency.**

SOUND ISOLATION: Vibration of Walls

In addition to a wall's simple to-and-fro motion, which depends on its weight or mass, a wall can have other forms of motion. Certain "favorite" frequencies for bending waves can exist, depending on the wall's stiffness. When these favorite waves occur, a material's resistance to sound is greatly reduced; this is called "coincidence" effect. The wave coincidence is shown by the exaggerated sketch below.

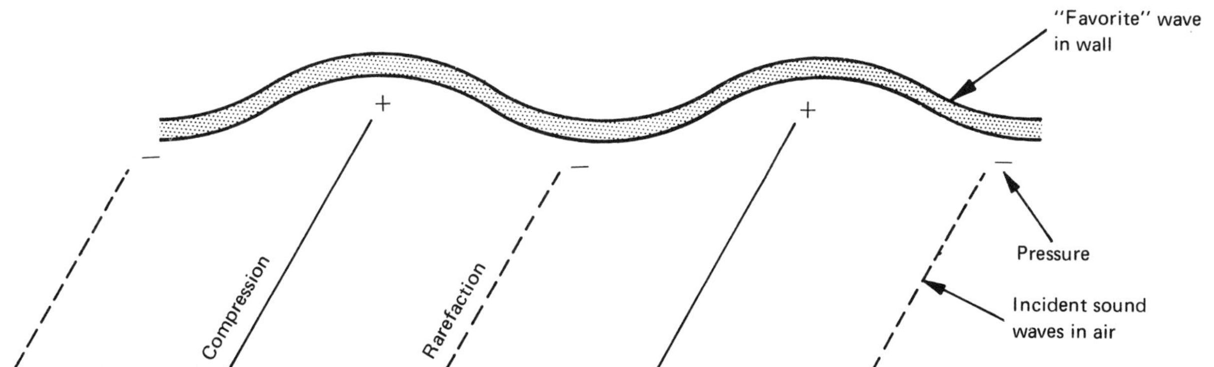

In practice, a wall's TL at certain frequencies will be much less than its mass alone would lead us to predict. Rather than a straight line, the TL curve for most single walls has the following shape:

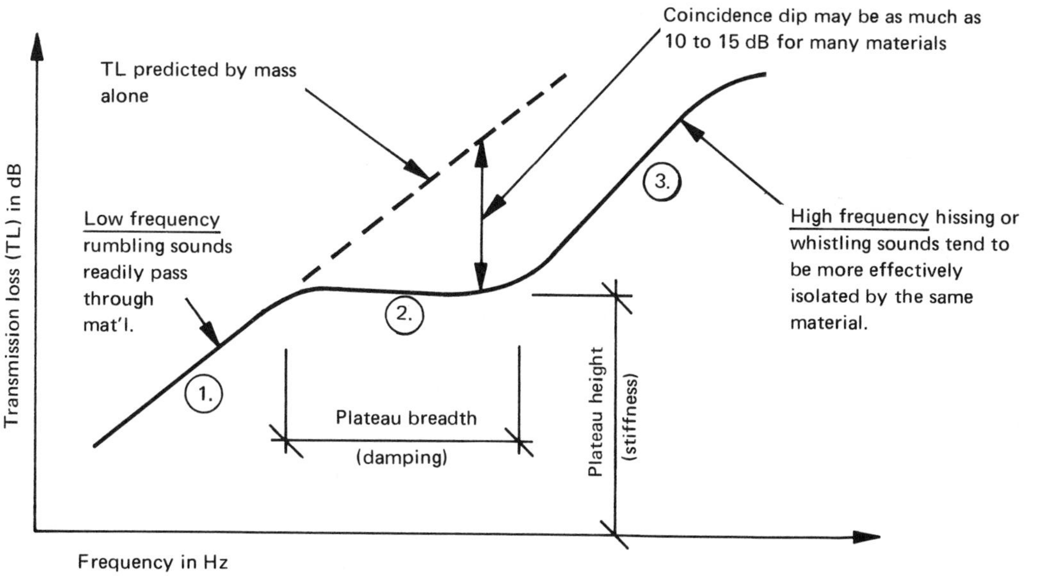

Typical TL curve shape for most single walls

The TL curve for most single walls consists of three basic parts or regions:

1. **Low frequency mass controlled region at 6 dB/octave slope**
2. **Constant plateau of TL region, which depends on bending stiffness and internal damping**
3. **Mass controlled region above plateau at 10 dB/octave slope**

NOTE: **Curves on the following page compare TL effectiveness for various materials of identical surface weight. (Page 82 is a design guide for estimating TL for common building materials.)**

SOUND ISOLATION: Transmission Loss for Some Homogeneous Materials (Equal Surface Weight in psf Assumed for All Materials)

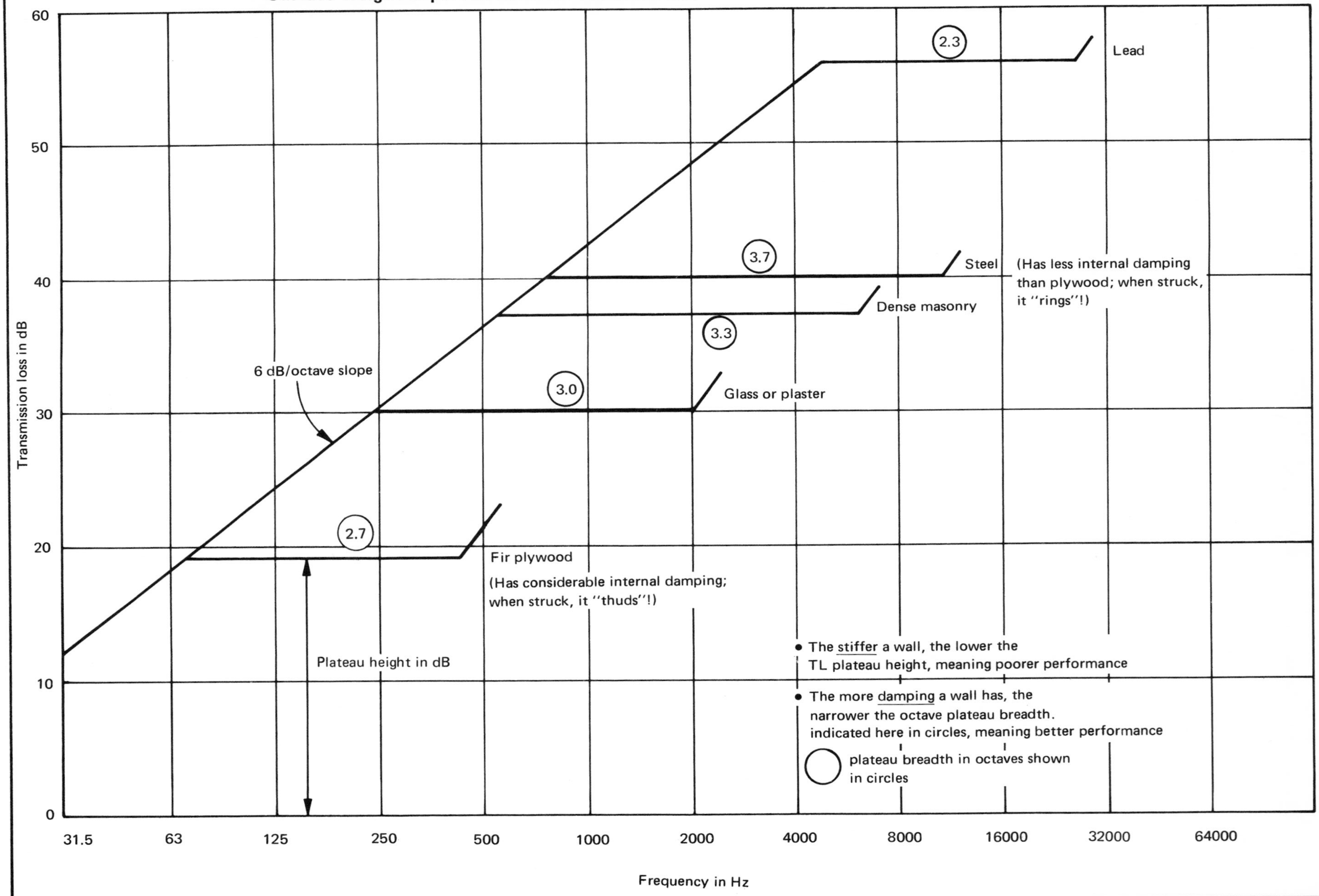

Transmission loss in dB

60
50
40
30
20
10
0

2.3
Lead

3.7
Steel
(Has less internal damping than plywood; when struck, it "rings"!)

Dense masonry
3.3

6 dB/octave slope

3.0
Glass or plaster

2.7
Fir plywood
(Has considerable internal damping; when struck, it "thuds"!)

Plateau height in dB

- The stiffer a wall, the lower the TL plateau height, meaning poorer performance
- The more damping a wall has, the narrower the octave plateau breadth. indicated here in circles, meaning better performance

◯ plateau breadth in octaves shown in circles

The more "limp" a material, the higher its plateau of TL.

31.5 | 63 | 125 | 250 | 500 | 1000 | 2000 | 4000 | 8000 | 16000 | 32000 | 64000

Frequency in Hz

SOUND ISOLATION: Mass Law

According to "mass law," a construction's STC rating* increases about 5 for each doubling of surface weight. STC data from a sample of typical building constructions show considerable scatter about the theoretical mass law curve. Accordingly, mass law is used only to establish a construction's TL at low frequencies below coincidence.

***STC (sound transmission class) is a single-number rating of a construction's TL performance over a standard frequency range. The higher the STC, the more efficient the construction for reducing sound transmission.** **See pages 69 through 71 for information on the STC rating method.**

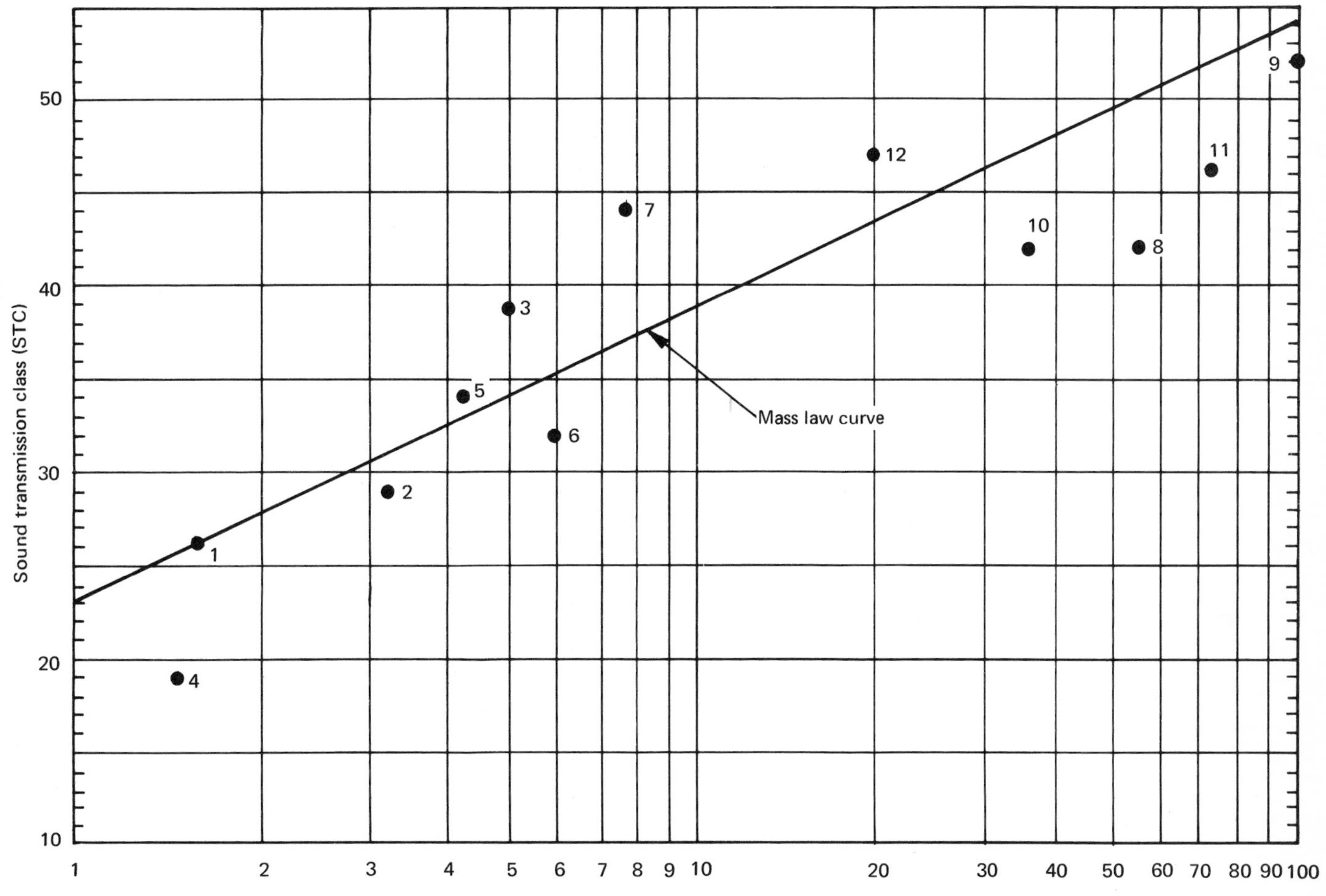

Building construction description:

- Glass
  1.) 1/8" single plate-glass pane
  2.) 1/4" single plate-glass pane
  3.) 1/4 & 1/8" double plate-glass window w/2" air space
- Doors
  4.) 1 3/4" hollow wood core door
  5.) 1 3/4" solid wood core door
- Walls
  6.) 2 x 4 wood studs w/1/2" gyp bd. both sides
  7.) 2 x 4 staggered wood studs w/1/2" gyp. bd. both sides
  8.) 4 1/2" brick w/1/2" gyp. plaster both sides
  9.) 9" brick w/1/2" gyp. plaster both sides
- Floor-ceilings
  10.) 3" concrete slab
  11.) 6" concrete slab
  12.) 18" steel joints w/1 5/8" concrete on 5/8" plywood and carpet on floor side. 5/8" gyp. bd. on ceiling side.

SOUND ISOLATION: TL Improvement from Doubling Wall Weight

For a barrier of dense concrete:

	Thickness, in.	Surface Wt., psf	STC Rating
A.	3	36	42
B.	6	72	46
C.	12	144	51
	24	288	58

Mass law follows a law of diminishing returns, as shown by the table above. The STC of a homogeneous construction increases by about 5 for each doubling of weight. However, it is the initial doubling that provides the most practical improvement since each successive doubling produces proportionally less improvement per unit weight, along with a greater increase in cost per unit STC (or TL) increase. Consequently, to achieve high TLs, double or complex constructions are often required.

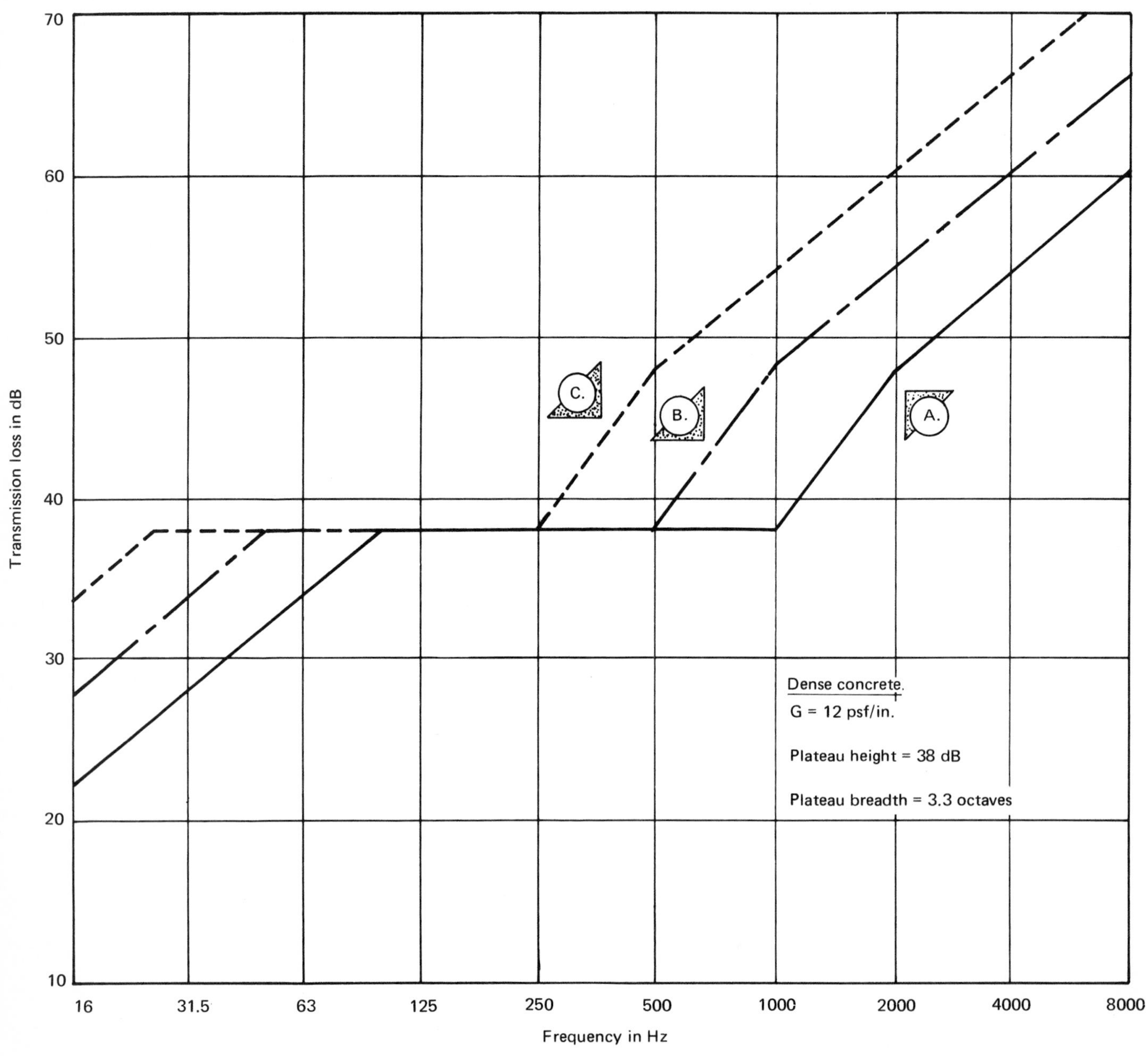

SOUND ISOLATION: Effect of Reduced Stiffness

A material's sound isolation efficiency depends on its stiffness as well as its mass!

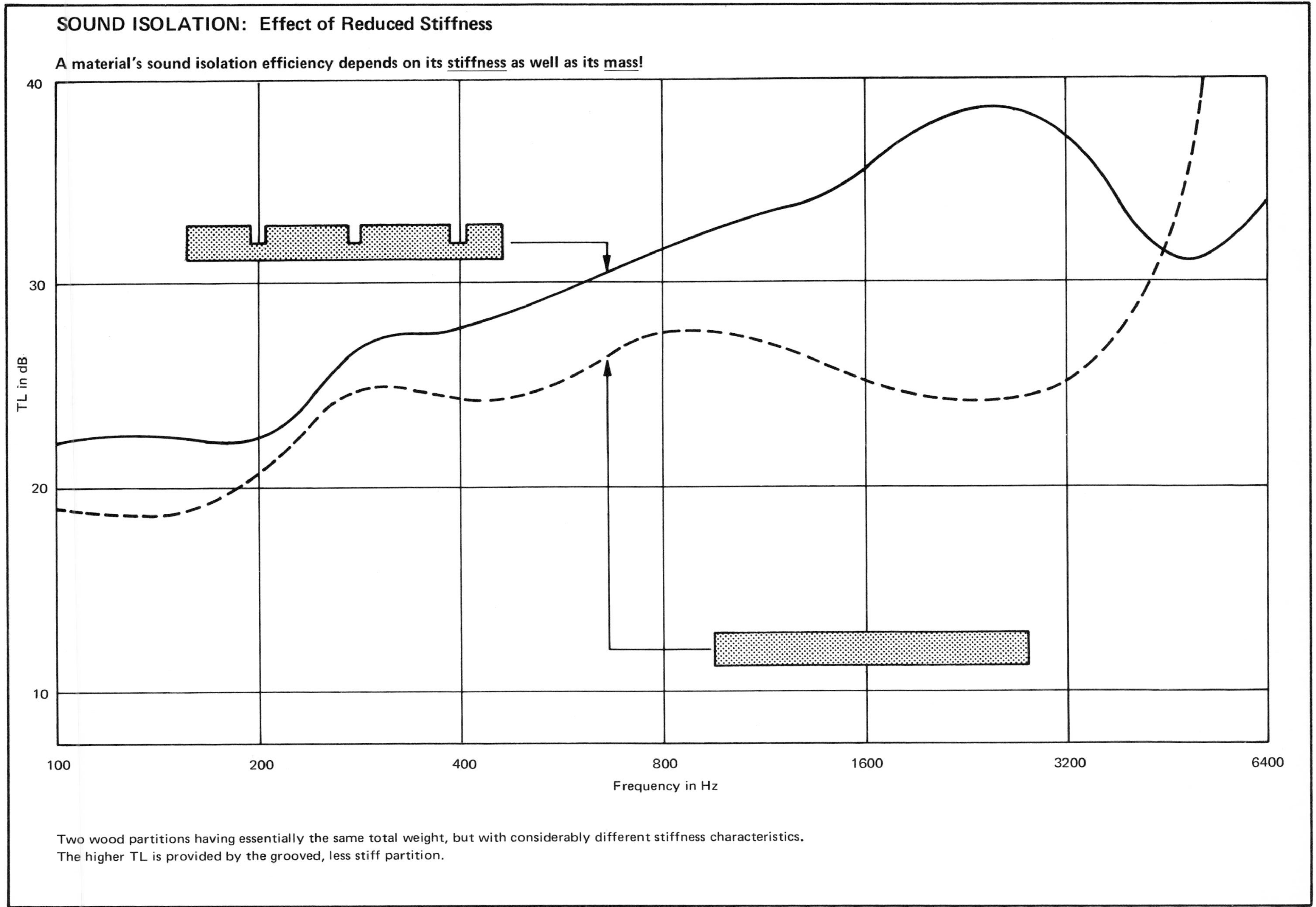

Two wood partitions having essentially the same total weight, but with considerably different stiffness characteristics.
The higher TL is provided by the grooved, less stiff partition.

SOUND ISOLATION: Sound Leaks

All sound leaks are important because sound will travel through any opening with little loss. The more effective the construction as a sound isolator, however, the more serious the leak, as shown by the curves below. Consequently, always avoid cracks, louvered doors, back-to-back electrical outlets, etc. For example, a 1 sq in. hole in a 100 sq ft of gypsum board partition can transmit as much sound as the rest of the partition!

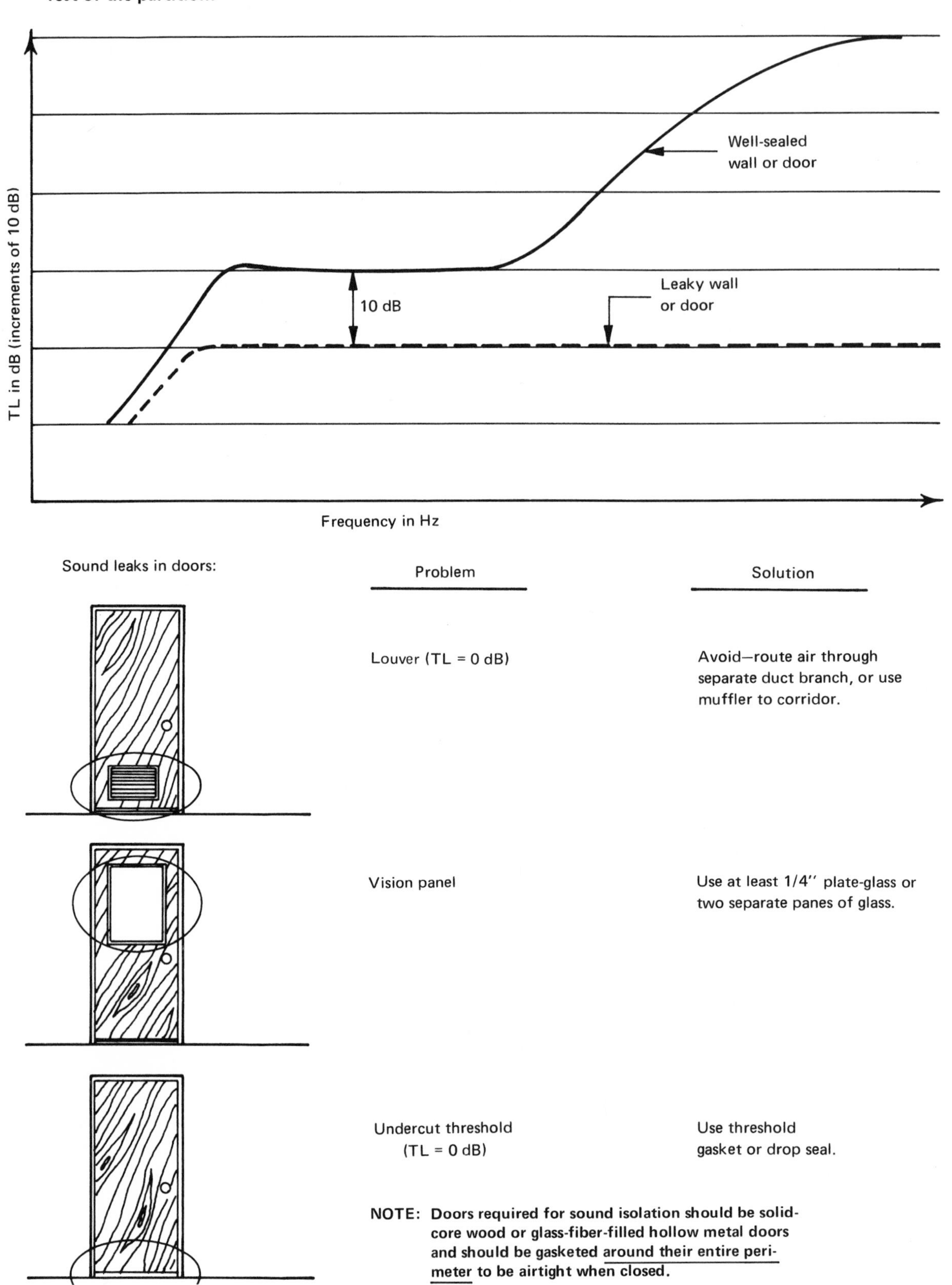

Problem	Solution
Louver (TL = 0 dB)	Avoid—route air through separate duct branch, or use muffler to corridor.
Vision panel	Use at least 1/4" plate-glass or two separate panes of glass.
Undercut threshold (TL = 0 dB)	Use threshold gasket or drop seal.

NOTE: Doors required for sound isolation should be solid-core wood or glass-fiber-filled hollow metal doors and should be gasketed around their entire perimeter to be airtight when closed.

SOUND ISOLATION: Composite Construction TL

When a "weaker" element, such as a window or door, is used in a construction, the composite TL for the combination is usually closer to the TL of the "weaker" element.

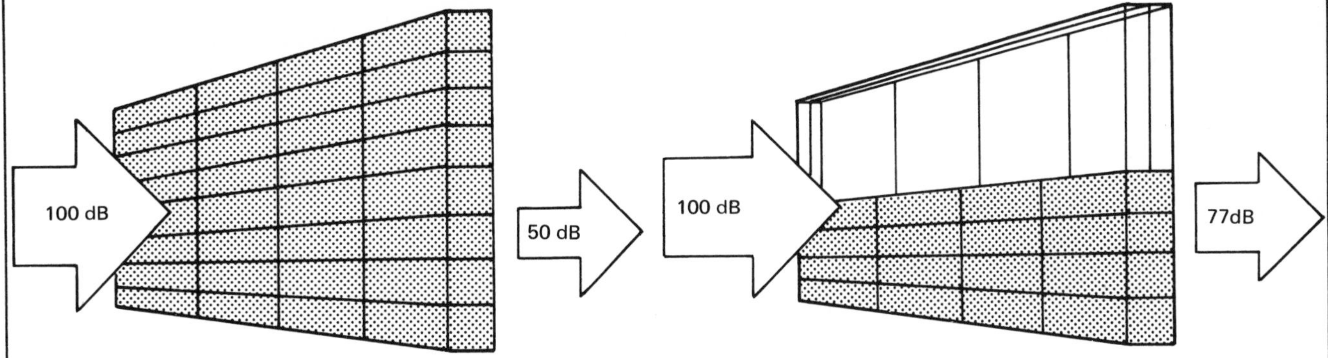

Case 1 - All brick
TL = 50 dB

Case 4 - 1/2 glass, 1/2 brick
TL = 23 dB

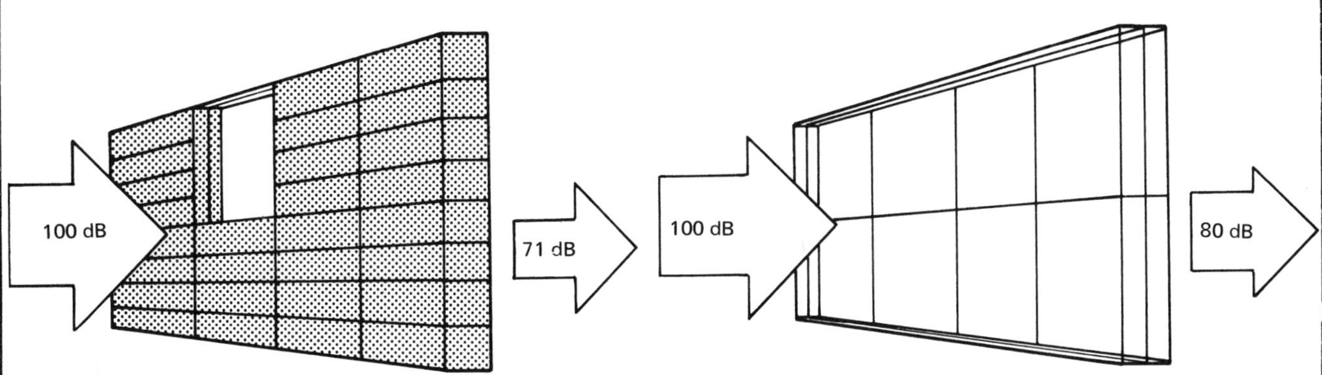

Case 2 - 1/8 glass, 7/8 brick
TL = 29 dB

Case 5 - All glass
TL = 20 dB

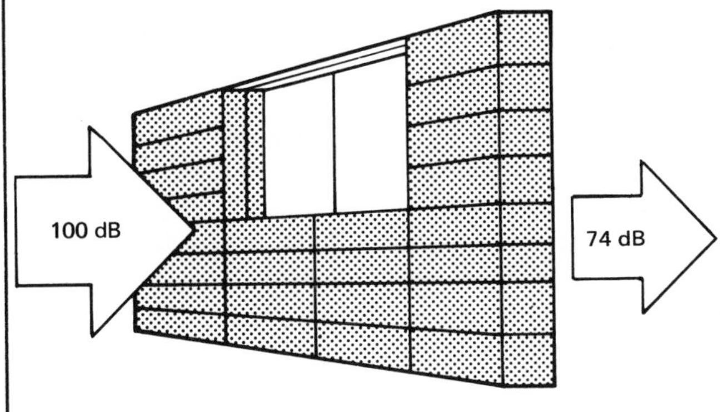

Case 3 - 1/4 glass, 3/4 brick
TL = 26 db

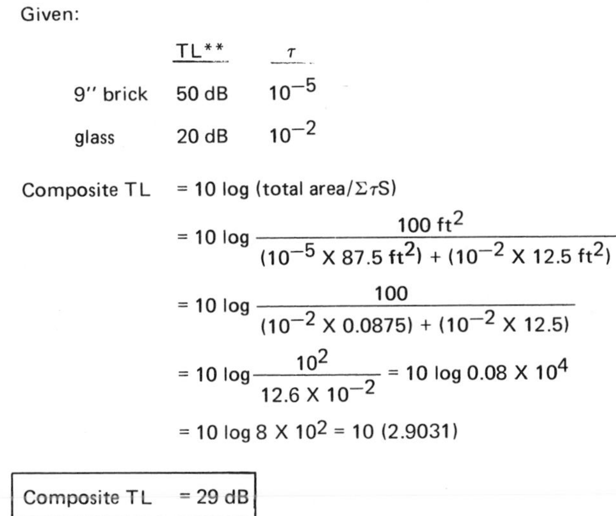

Example - Case 2:*

Given:

	TL**	τ
9" brick	50 dB	10^{-5}
glass	20 dB	10^{-2}

Composite TL $= 10 \log (\text{total area}/\Sigma\tau S)$

$$= 10 \log \frac{100 \text{ ft}^2}{(10^{-5} \times 87.5 \text{ ft}^2) + (10^{-2} \times 12.5 \text{ ft}^2)}$$

$$= 10 \log \frac{100}{(10^{-2} \times 0.0875) + (10^{-2} \times 12.5)}$$

$$= 10 \log \frac{10^2}{12.6 \times 10^{-2}} = 10 \log 0.08 \times 10^4$$

$$= 10 \log 8 \times 10^2 = 10\ (2.9031)$$

Composite TL = 29 dB

*See also next page for solving case 2 with a chart.

**TL = 10 log $1/\tau$

SOUND ISOLATION: Chart for Determining the Transmission Loss of Composite Structures

Alternative solution to Case 2 from preceding page:

With TL $_{brick}$ – TL $_{glass}$ = 50 – 20 = <u>30</u> dB and 1/8 glass (about <u>13%</u>)
Read <u>21</u> dB from bottom scale (see dashed lines)
Therefore, Comp. TL = 50 – 21 = 29 dB

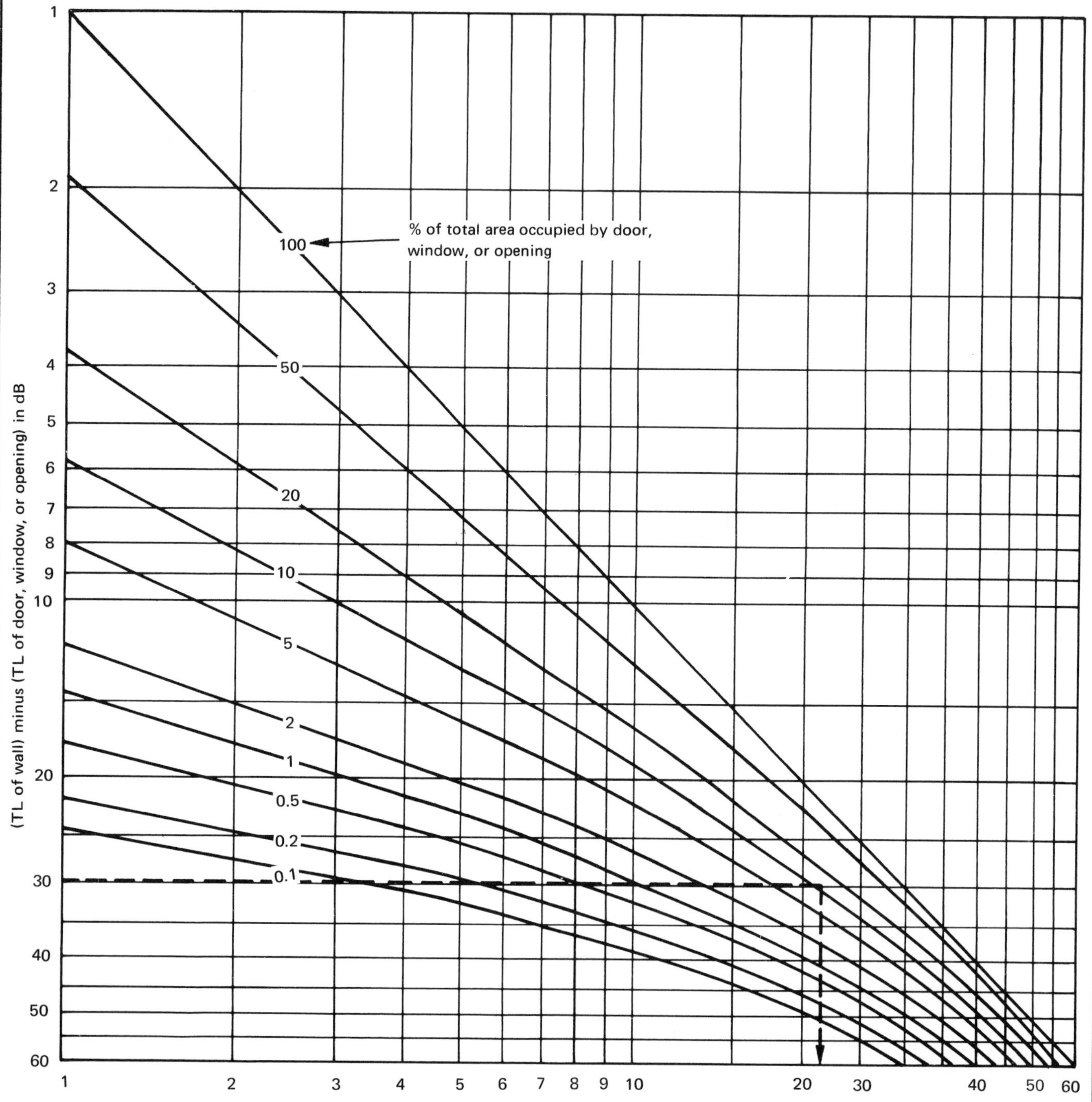

NOTE: **The following general guidelines may be used to plan composite constructions. If the % of a "weaker" element is less than 25%, its TL may be 5 dB lower than the wall TL; if 25 to 50%, it may be 2 dB lower; and if greater than 50%, the composite TL will be equai to that of the "weaker" element. Actual composite TLs should be found from the above chart or by the formula:**

$$\text{Composite TL} = 10 \log (\text{tot. area}/\Sigma \tau S)$$

NOISE REDUCTION BETWEEN ROOMS

Transmission loss (TL) and noise reduction (NR) between rooms are two concepts basic to all sound isolation problems. Both TL and NR are expressed in decibels. The NR between rooms is simply the arithmetical difference in room intensity levels, that is, the noise in the source room at an intensity level of IL_1 less the transmitted noise in the receiving room at a reduced intensity level of IL_2.

$$NR = IL_1 - IL_2 \qquad \text{(in dB at a given frequency)}$$

Our ears sense the NR or IL difference between rooms. NR at a given frequency is independent of source room noise level. For example, if the NR is 40 dB and IL_1 is 96 dB, IL_2 will be 56 dB. If IL_1 is 62 dB in this example, IL_2 will be 22 dB. NR is dependent on the three basic factors listed below:

1. **Area of Wall Transmitting Sound, S, in Square Feet:** The actual barrier size is important since it is the sound source in the receiving room. When a sound wave strikes the "front" side of a wall, its energy causes the whole wall to vibrate. This vibration sets into oscillation the air particles along its "back" or opposite side. These vibrating air particles then radiate sound into the space on the back side of the wall.
2. **Absorption in the Receiving Room, a_2, in Sabins:** The noise buildup is greater in reverberant rooms than in "dead" or absorptive rooms.
3. **Transmission Loss of the Wall Common to Both Rooms, TL, in Decibels:** TL is a physical property obtained from laboratory tests. However, in practice TL varies with edge conditions, barrier size, and workmanship. Sloppy construction and "leaky" walls can ruin any TL prediction. Typically, TLs may range between about 10 and 80 dB.

The above factors are related to NR by the following formula:

$$NR = TL + 10 \log \frac{a_2}{S} \qquad \text{(in dB at a given frequency)}$$

Note that the factor $10 \log (a_2/S)$, which may range between about +6 and −6 dB, can be estimated from the graph on page 67. If the barrier size in square feet coincidentally equals the sabins in the receiving room (the situation in many small enclosed office spaces with absorptive ceilings), the quantity $10 \log (a_2/S)$ will equal 0 because $10 \log 1 = 0$. Consequently NR will equal TL; however, for most situations the NR will be either higher or lower than the TL of the barrier, as shown on page 66. Also, if the common wall is constructed of two or more materials (e.g., with a window, door, etc.), the "composite TL" should be substituted for TL in the above formula.

Example: Two adjacent apartment living rooms have a common partition of 4-in. brick. The wall area S is 200 sq ft. Assume a TL of 40 dB for the wall at 500 Hz. Both rooms have 300 sabins of absorption a_2 at 500 Hz. If the intensity level IL_1 is 82 dB in room 1 from a loud stereo, what is IL_2 heard in room 2?

$$NR = TL + 10 \log \frac{a_2}{S}$$

$$NR = 40 + 10 \log \frac{300}{200} = 40 + 1.8 = 41.8 \text{ dB at 500 Hz}$$

and $\quad IL_2 = IL_1 - NR$

$$IL_2 = 82 - 41.8 = 40.2 \simeq 40 \text{ dB at 500 Hz in room 2}$$

SOUND ISOLATION: Transmission Loss and Noise Reduction Between Rooms

The TL is a material property that cannot be viewed as a unique value for a given wall construction for all conditions. The examples below demonstrate that the same partition is more effective when transmitting sound into a large, highly absorbent receiving room than into a small, reverberant receiving room.

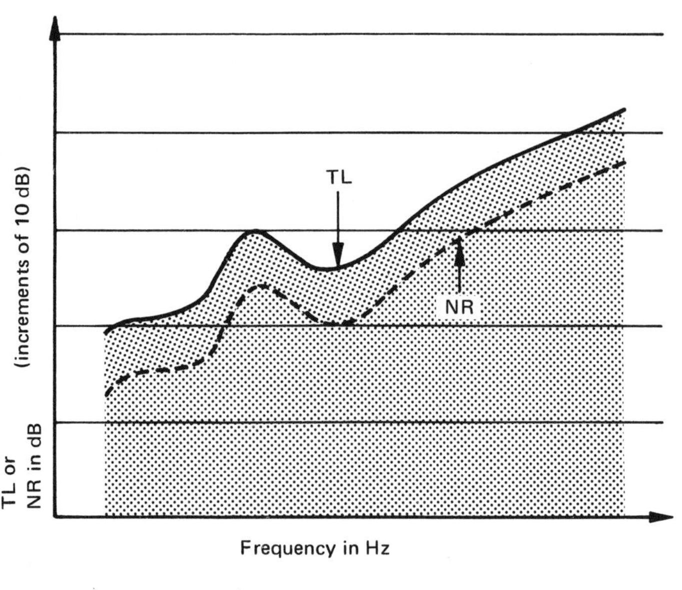

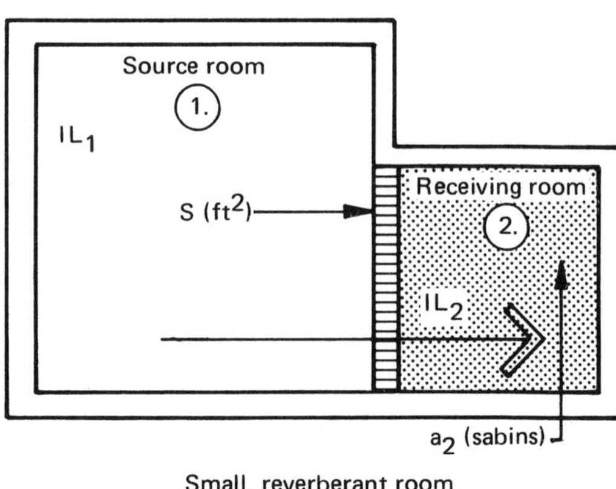

Small, reverberant room

NOTE: **The TL curves shown on both graphs are identical, as same partition is used in both examples.**

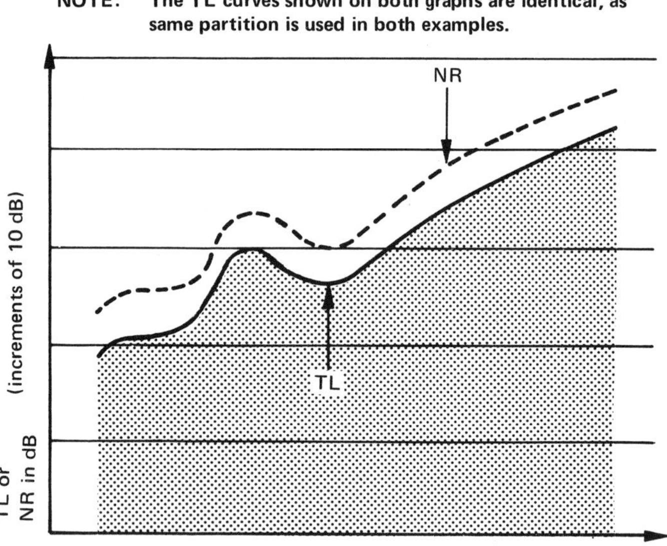

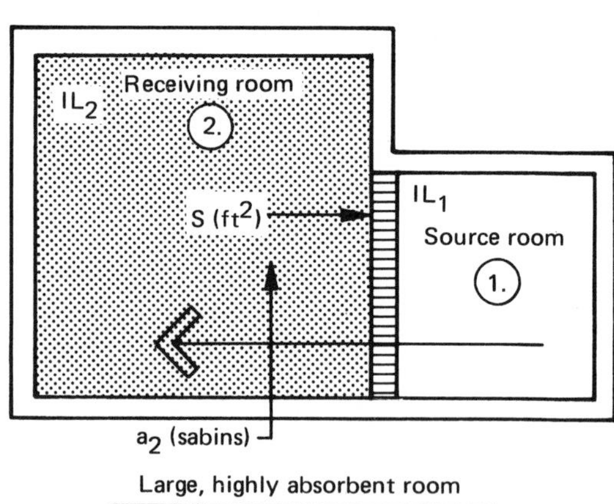

Large, highly absorbent room

Noise reduction between rooms is the actual difference in intensity levels between the source room and receiving room, that is NR = $IL_1 - IL_2$ in dB. Also, NR between rooms can be predicted by

$$NR = TL + 10 \log a_2/S$$

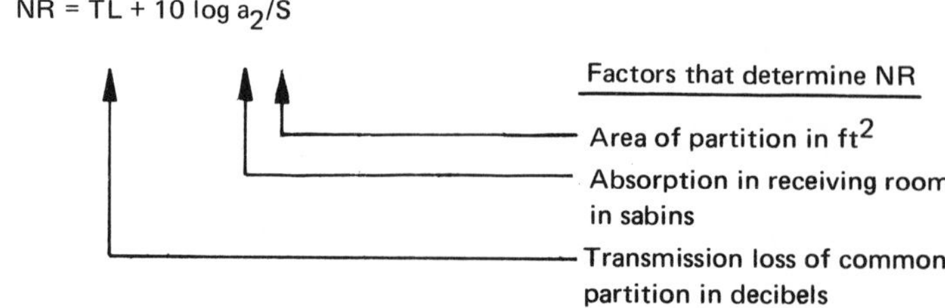

SOUND ISOLATION: Noise Reduction–Receiving Room Absorption and Wall Area Graph

$NR = TL + \boxed{10 \log a_2/S}$ (in dB at a given frequency)

Graph below solves this term.

Example - use of graph:

Given: 50 ft^2 wall area (S) with TL of 40 dB and 150 sabins of receiving room absorption (a_2), both at 250 Hz.

Procedure to find NR:

1. Enter graph at a_2 = 150 sabins and read opposite 50 ft^2 curve to $\simeq$5 dB (see dashed lines)
2. Therefore, NR = TL + 5 and NR = 40 + 5 = 45 dB at 250 Hz

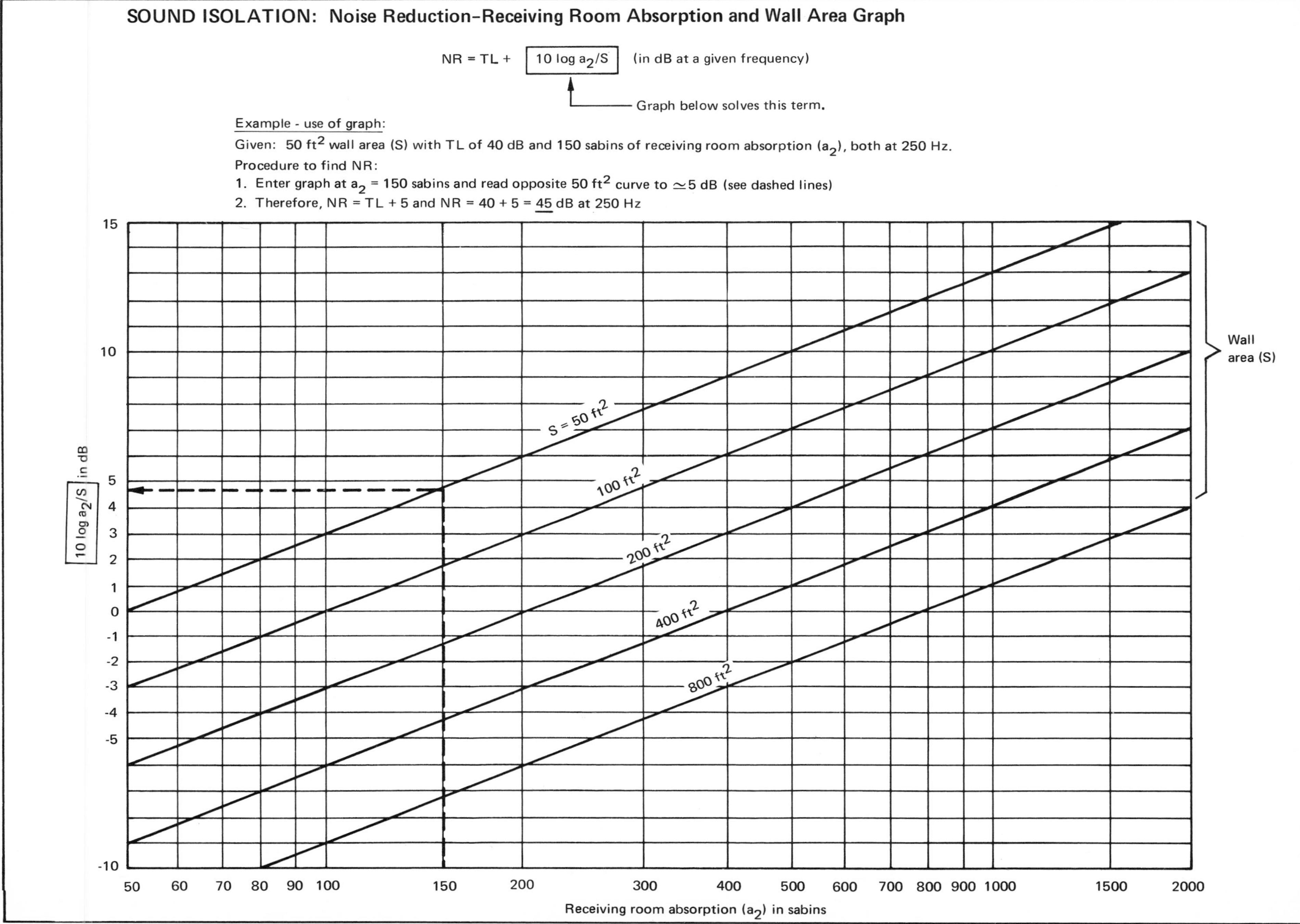

SOUND ISOLATION: Typical TL Test Rooms

The "TL" test panel shown below covers an opening between two rooms constructed with thick, massive walls which transmit much less sound energy than the test panel. Consequently, essentially all the transmission between the rooms can be considered to take place through the test panel.

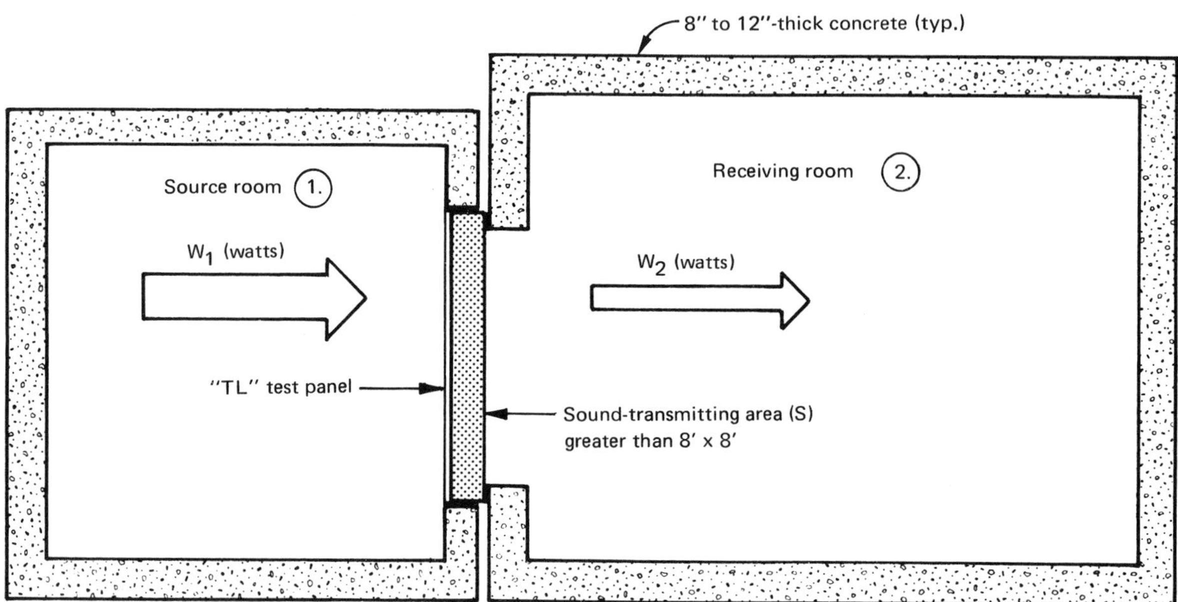

TL TEST REFERENCE STANDARDS:

1. ASTM Standard E90-70, "Laboratory Measurement of Airborne Sound Transmission Loss of Building Partitions."
2. ISO Recommendation R140, "Field and Laboratory Measurements of Airborne and Impact Sound Transmission," 1960.

TL FORMULAS:

1. TL = 10 log $1/\tau$, where τ = sound transmission coefficient (dimensionless).
2. $\tau = W_2/W_1$, where W_2 = radiated sound power, watts; W_1 incident sound power, watts.

SOUND ISOLATION: Sound Transmission Class (STC) Rating System Examples

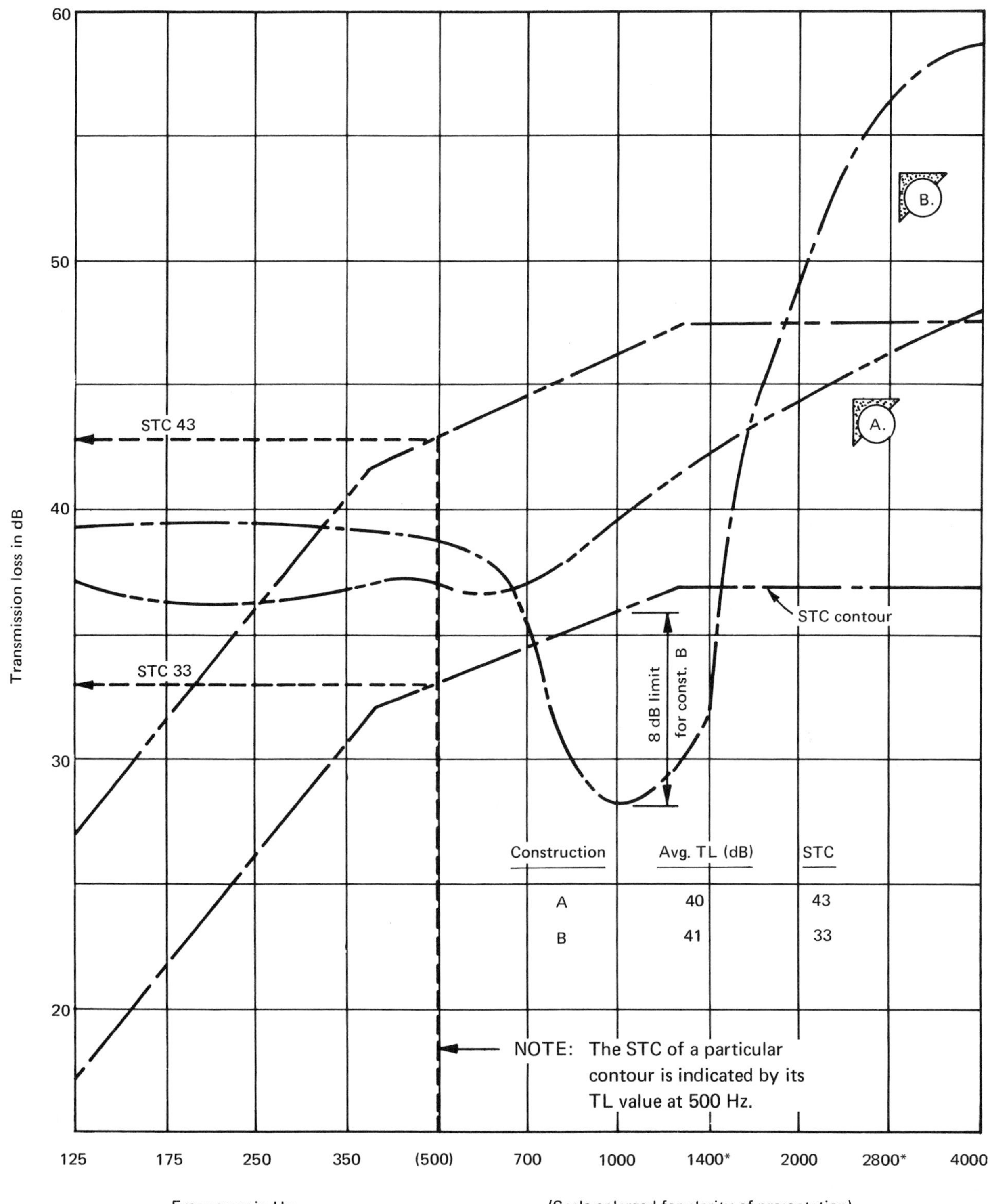

Until 1961 the nine-frequency average (*values above not used in computing nine-frequency average) was used to establish a TL rating for a construction. This system was undesirable because it implied that good isolation at one frequency could overcome poor isolation at another. The STC system rates the entire TL curve, using a standard contour as the reference. The contour is shown above at its rating positions for example constructions A and B.

SOUND ISOLATION: Sound Transmission Class (STC) Overlay Contour and Grid

Trace the STC contour and grid shown below on a transparent overlay which can then be used to determine STC ratings according to ASTM procedures outlined on the following page.

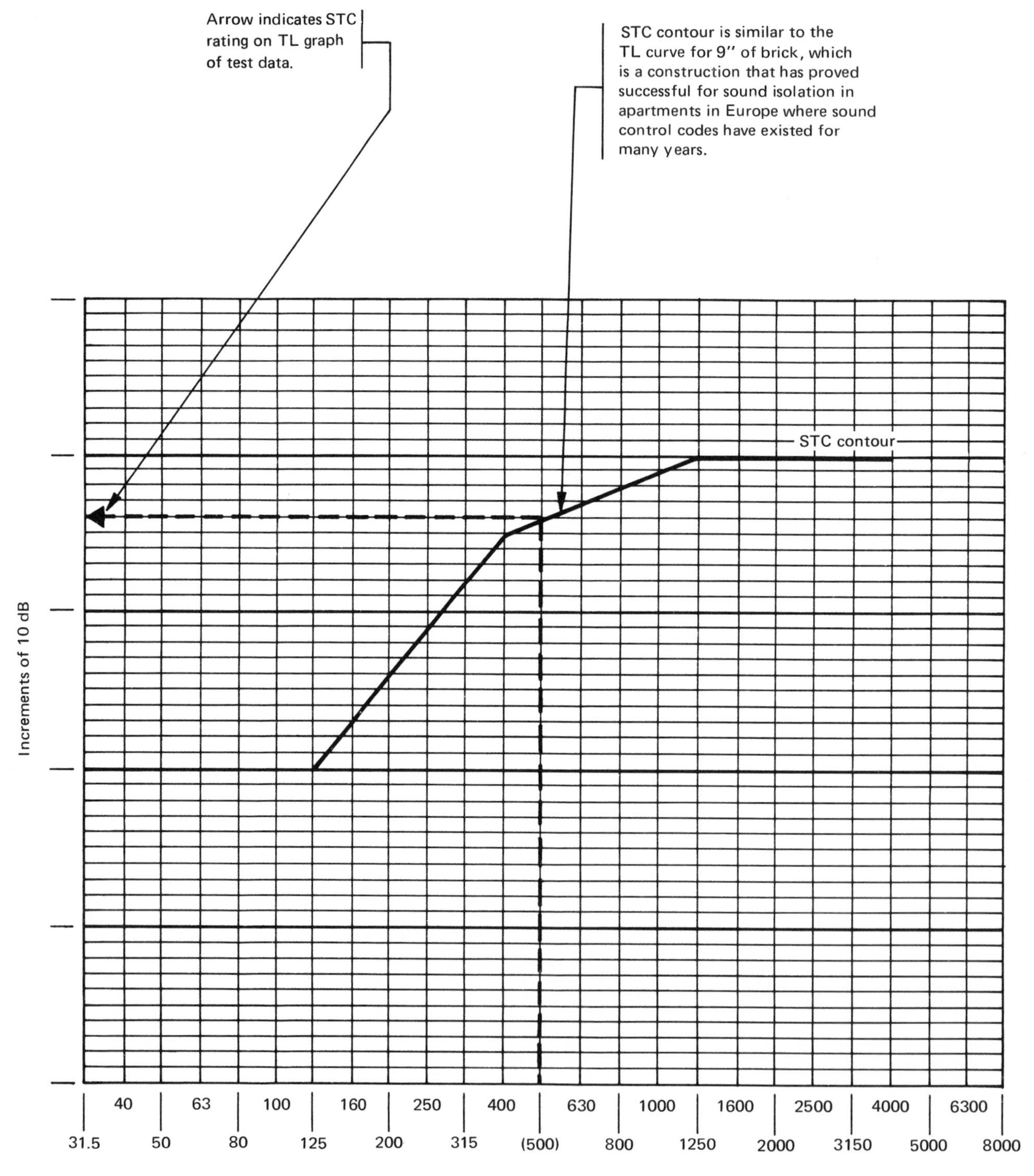

SOUND TRANSMISSION CLASS (STC) RATING PROCEDURES

The STC is a single-number rating of a construction's airborne sound transmission performance at different frequencies. The higher the STC rating, the more efficient the construction for reducing sound transmission in the test frequency range.

The STC rating method procedures are specified in the American Society for Testing and Materials (ASTM) annual book of standards. The sound transmission loss (TL) of a construction's test specimen is measured at 16 third-octave bands with center frequencies from 125 to 4000 Hz. To determine the STC of a given specimen, its measured TL values are plotted against frequency and compared with a reference STC contour. The STC rating is easily determined by using a transparent overlay on which the STC contour is drawn, as shown on the preceding page. The STC contour is shifted vertically relative to the test data curve to as high a position as possible according to the following conditions:

1. The maximum deviation of the test curve below the contour at any single test frequency shall not exceed 8 dB.
2. The sum of the deviations at all 16 frequencies of the test curve below the contour shall not exceed 32 dB – an average deviation of 2 dB.

When the STC contour is adjusted to the highest position that meets the above requirements, the STC rating is read from the vertical scale of the test curve as the TL value corresponding to the intersection of the STC contour and the 500 Hz ordinate. In the examples on page 69, the STC ratings reflect the significance of deficiencies such as the dip in construction B's TL curve. B's STC 33 rating, indicated on the graph, is governed by the 8 dB deviation below the contour at 1000 Hz, although its total deviation is well under 32 dB. Construction A has an STC 43 rating, also shown on the graph.

NOTE: Results from a simplified method to determine the field STC directly from single A-scale measurements on the source and receiving sides of test constructions seem to agree with ASTM E336-70T results in most situations. For a detailed description of the method, see Siekman, W.; Yerges, J. F.; and Yerges, L. F.; A Simplified Field Sound Transmission Test, *Sound and Vibration*, Vol. 5, No. 10 (October 1971).

References

1. "Laboratory Measurement of Airborne Sound Transmission Loss of Building Partitions," ASTM Designation E90-70.
2. "Measurement of Airborne Sound Insulation in Buildings," ASTM Designation E336-70T.
3. "Determination of Sound Transmission Class," ASTM Designation E413-70T.

SOUND ISOLATION: Ideal vs. Real Conditions

Lab, and field measurements of a 6″ masonry block wall, painted, are shown below.

Lab.

Field

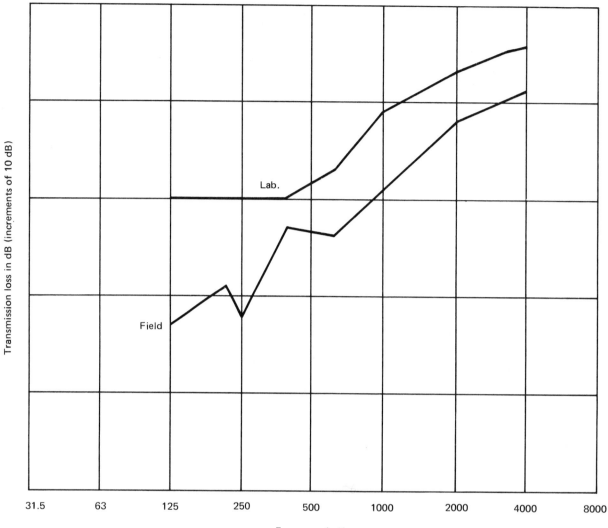

Transmission loss in dB (increments of 10 dB)

31.5 63 125 250 500 1000 2000 4000 8000

Frequency in Hz

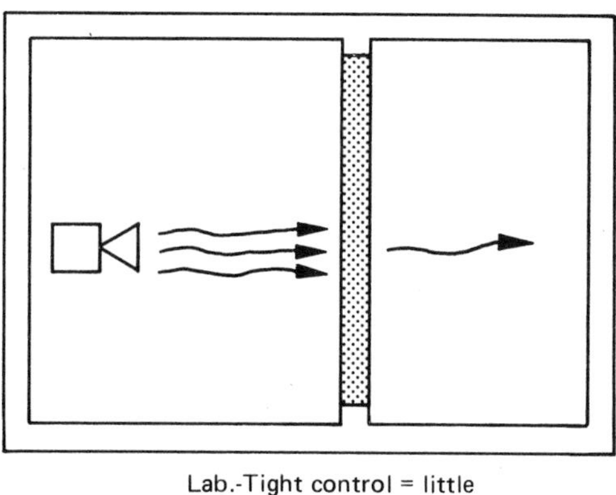

Lab.-Tight control = little or no flanking

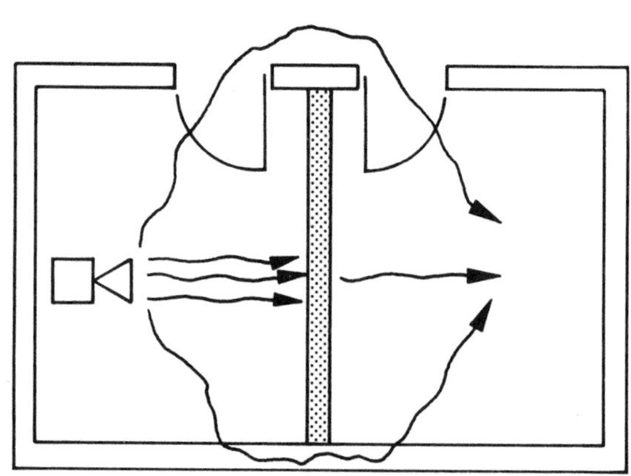

Field-Loose control = almost always flanking

Lab. data generally reflect idealized conditions. Therefore, it is important to account for leaks and flanking paths which reduce performance of barriers when they are installed in buildings. Use balanced design, supervise construction closely, and consider all possible flanking paths.

TRANSMISSION LOSS DATA FOR COMMON BUILDING CONSTRUCTIONS*

Building Construction	Transmission Loss, dB						STC Rating	IIC Rating†
	125 Hz	250 Hz	500 Hz	1000 Hz	2000 Hz	4000 Hz		
Walls(1, 3, 4)								
Interior:								
1. 2-in. solid plaster on metal lath (18 psf)	20	22	22	27	36	42	24	
2. 2 x 4 wood studs 16 in. o. c. with 1/2-in. gypsum board both sides (6 psf)	10	28	33	42	47	41	32	
3. 2 x 4 staggered wood studs 16 in. o. c. each side with 1/2-in. gypsum board both sides (7.7 psf)	23	32	40	51	53	44	41	
4. Construction No. 3 with 1 1/2 in. glass-fiber insulation in cavity (7.8 psf)	27	38	46	54	57	52	49	
5. 2 x 4 wood studs 16 in. o. c. with 5/8-in. gypsum board both sides – one side screwed to resilient channels. 3 in. glass-fiber batt insulation in cavity (6.5 psf)	32	42	52	58	53	54	52	
6. 6-in. concrete block wall, painted (34 psf)	37	36	42	49	55	58	44	
7. 8-in. concrete block wall with 3/4-in. wood furring, gypsum lath and plaster both sides (67 psf)	43	47	47	55	58	60	52	
8. 2 1/2-in. steel channel studs with 5/8-in. gypsum board both sides (11 psf)	15	24	38	48	40	42	36	
9. Construction No. 8 with glass-fiber insulation in cavity	23	35	44	53	45	43	39	
10. 2 5/8-in steel channel studs with two layers 5/8-in. gypsum board one side, 1 layer other side (8 psf)	22	26	40	51	44	47	40	
11. Construction No. 10 with glass-fiber insulation in cavity	28	38	50	57	50	50	46	
12. 2 5/8-in. steel channel studs with 2 layers 5/8-in. gypsum board both sides (10 psf)	27	34	48	55	50	57	45	

*Manufacturer's TL data for proprietary building constructions should be from up-to-date full-scale tests by acoustical laboratories (e.g., Cedar Knolls, Geiger and Hamme, Riverbank, etc.) or from actual field tests on their installed products. For example, the performance of movable and operable wall systems can be significantly degraded by poor gasketing seals and closures in the actual installation, or by flanking paths due to not using balanced designs.

†See p. 74

Building Construction	Transmission Loss, dB 125 Hz	250 Hz	500 Hz	1000 Hz	2000 Hz	4000 Hz	STC Rating	IIC Rating†
13. Construction No. 12 with glass-fiber insulation in cavity	33	44	55	60	55	60	55	
Exterior:								
14. 4 1/2-in. brick with 1/2-in. plaster each side (55 psf)	34	34	41	50	56	58	42	
15. 9-in. brick with 1/2-in. plaster each side (100 psf)	41	43	49	55	57	60	52	
16. Two wythes of plastered 4 1/2-in. brick, 2-in. air space with glass-fiber in cavity (90 psf)	43	50	52	61	73	78	59	
17. 2 x 4 wood studs 16 in. o. c. with 1-in. stucco on metal lath on outside and 1/2-in. gypsum board on inside (8 psf)	21	33	41	46	47	51	42	
18. 6-in. concrete with 1/2-in. plaster both sides (80 psf)	39	42	50	58	64	67	53	
Floor-Ceilings(1)								
19. 2 x 10 wood joists 16 in. o. c. with 1/2-in. plywood subfloor, 25/32-in. oak on floor side, and 5/8-in. gypsum board on ceiling side (10 psf)	23	32	36	45	49	56	37	32
20. Construction No. 19 with 3 in. glass-fiber batt insulation in cavity	25	36	38	46	51	57	40	32
21. Construction No. 19 with 5/8-in. gypsum board screwed to resilient channels, spaced 24 in. o. c. perpendicular to joists	30	35	44	50	54	60	47	39
22. 4-in. thick reinforced concrete slab (53 psf)	48	42	45	56	58	66	44	25
23. 6-in. thick reinforced concrete slab with 3/4-in. T & G wood flooring on 1 1/2 x 2 in. wooden battens on 1-in. thick glass-wool quilt (83 psf)	38	44	52	55	60	65	55	57
24. 18-in. steel joists 16 in. o. c. with 1 5/8-in. concrete on 5/8-in. plywood nailed to joists and heavy carpet on underlay. On ceiling side, 5/8-in. gypsum board nailed to joists (20 psf)	27	37	45	54	60	65	47	62

†IIC (impact insulation class) is a single-number rating of a floor-ceiling construction's impact sound transmission performance over a standard frequency range. The higher the IIC rating, the more efficient the construction for reducing impact sound transmission. See page 136 for a discussion of IIC rating procedures.

Building Construction	Transmission Loss, dB 125 Hz	250 Hz	500 Hz	1000 Hz	2000 Hz	4000 Hz	STC Rating	IIC Rating†
Roofs(1)								
25. 3 x 8 wood beams 32 in. o. c. with 2 x 6 T & G planks, asphalt felt built-up roofing and gravel topping	29	33	37	44	55	63	43	
26. Construction No. 25 with 2 x 4s 16 in. o. c. between beams with 1/2-in. gypsum board supported by metal channels on ceiling side, and 4-in glass-fiber batt insulation in cavity	35	42	49	62	67	79	53	
27. Corrugated steel, 24-gauge with 1 3/8-in. sprayed cellulose insulation on ceiling side (1.8 psf)	17	22	26	30	35	41	30	
28. 2 1/2-in. sand gravel concrete (148 pcf) on 28-gauge corrugated steel supported by 14-in steel bar joists, with 1/2-in. gypsum plaster on metal lath and 3/4-in. metal furring channels 13 1/2-in. o. c. on ceiling side (41 psf)	32	46	45	50	57	61	49	
Doors(1)								
29. Louvered door, 25 to 30% open area	10	12	12	12	12	11	12	
30. 1 3/4-in. hollow wood core door, no gaskets or closure, 1/4-in. air gap at sill	14	19	23	18	17	21	19	
31. Construction No. 30 with gaskets and drop seal	19	22	25	19	20	29	21	
32. 1 3/4-in. solid wood core door with gaskets and drop seal (4.3 psf)	29	31	31	31	39	43	34	
33. 1 3/4-in. hollow 16-gauge steel door, glass fiber filled core with gaskets and drop seal (6.8 psf)	23	28	36	41	39	44	38	
Glass(2)								
34. 1/8-in. single plate-glass pane	18	21	26	31	33	22	26	
35. 1/4-in. single plate-glass pane with rubber gasket	25	28	30	34	24	35	29	
36. 9/32-in. laminated glass pane (i.e., viscoelastic layer sandwiched between glass layers)	26	29	33	36	35	39	36	
37. 1/4 + 1/8-in. double plate-glass window with 2-in. air space	18	31	35	42	44	44	39	
38. Construction No. 37 with 4-in. air space	21	32	42	48	48	44	43	

Sources

1. Berendt, R. D. et al.: "*A Guide to Airborne, Impact, and Structure Borne Noise-Control in Multifamily Dwellings*," U. S. Department of Housing and Urban Development, September, 1967.
2. Bishop, D. E., and P. W. Hirtle: Notes on the Sound Transmission Loss of Residential-type Windows and Doors, *J. Acoust. Soc. Amer.*, Vol. 43, no. 4, April, 1968.
3. Newman, R. B., and W. J. Cavanaugh: Acoustics in J. H. Callender (ed.), "*Time-saver Standards*," 4th ed., McGraw-Hill, New York, 1966.
4. "Solutions to Noise Control Problems in the Construction of Houses, Apartments, Motels and Hotels," Owens-Corning Fiberglas Corp., Toledo, Ohio, 1965 and 1969 editions.

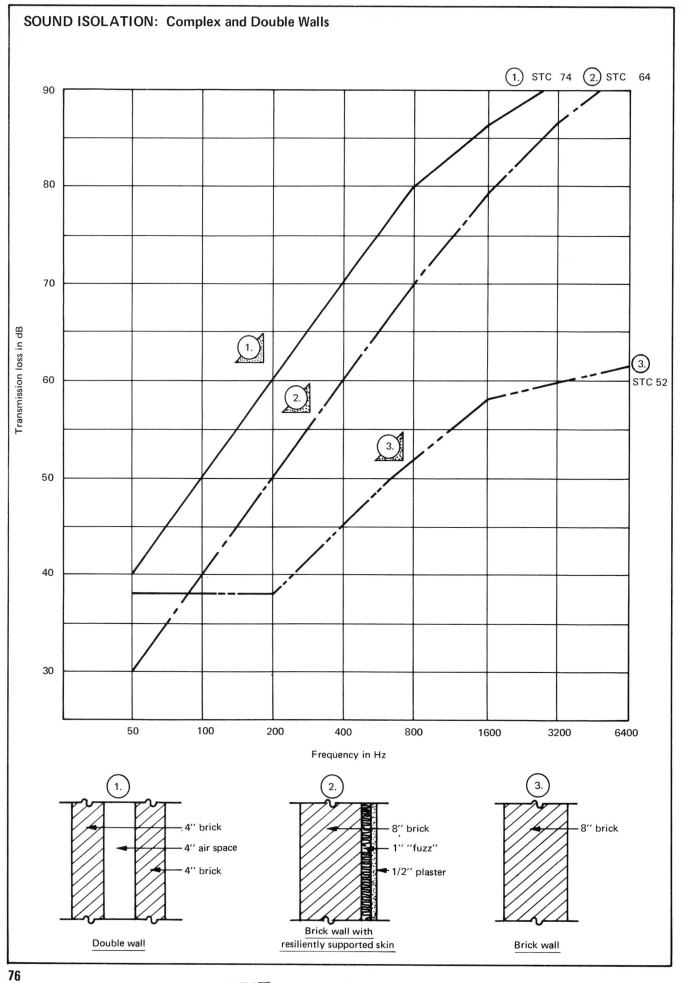

SOUND ISOLATION: Complex and Double Walls

1. STC 74
2. STC 64
3. STC 52

Transmission loss in dB: 30, 40, 50, 60, 70, 80, 90

Frequency in Hz: 50, 100, 200, 400, 800, 1600, 3200, 6400

1.
- 4″ brick
- 4″ air space
- 4″ brick

Double wall

2.
- 8″ brick
- 1″ "fuzz"
- 1/2″ plaster

Brick wall with resiliently supported skin

3.
- 8″ brick

Brick wall

DOUBLE WALL CONSTRUCTION TL IMPROVEMENTS

The TL of a wall can be greatly improved if the wall is made up of two or more layers separated by an air space. If the layers are separated by a considerable distance, they in effect form two independent walls, and the total TL can approach the sum of the individual panel TLs. For most practical situations, however, separations of 12 in. or less are more common, and the TL of the double wall construction is far less than the sum of the individual panel TLs. In all double wall construction rigid ties between panels must be avoided. The following table can be used to estimate the improvement in TL from splitting a wall of a given weight into two separate walls.

IMPROVEMENT IN TL, dB

Air space, in.	125 Hz	250 Hz	500 Hz	1000 Hz	2000 Hz	4000 Hz	Approx. STC Improve.
1 1/2	0	1	2	5	9	12	3
3	1	2	7	10	14	15	6
6	5	7	11	15	19	20	8

Example: Estimate the improvement in TL of a wall consisting of two 6-in. thick concrete-block layers, separated by a 6-in. air space, over a single 12-in. thick concrete-block wall.

TRANSMISSION LOSS, dB

Construction	125 Hz	250 Hz	500 Hz	1000 Hz	2000 Hz	4000 Hz	STC Rating
12-in. painted concrete-block wall with sand-filled cells (100 psf)	33	40	47	53	59	65	51
Improvement in TL (and STC)	5	7	11	15	19	20	8
6-in. concrete block + 6-in. air space + 6-in. concrete block (total weight = 100 psf)	38	47	58	68	78	85	59

SOUND ISOLATION: Transmission Loss of Floor-ceiling Constructions

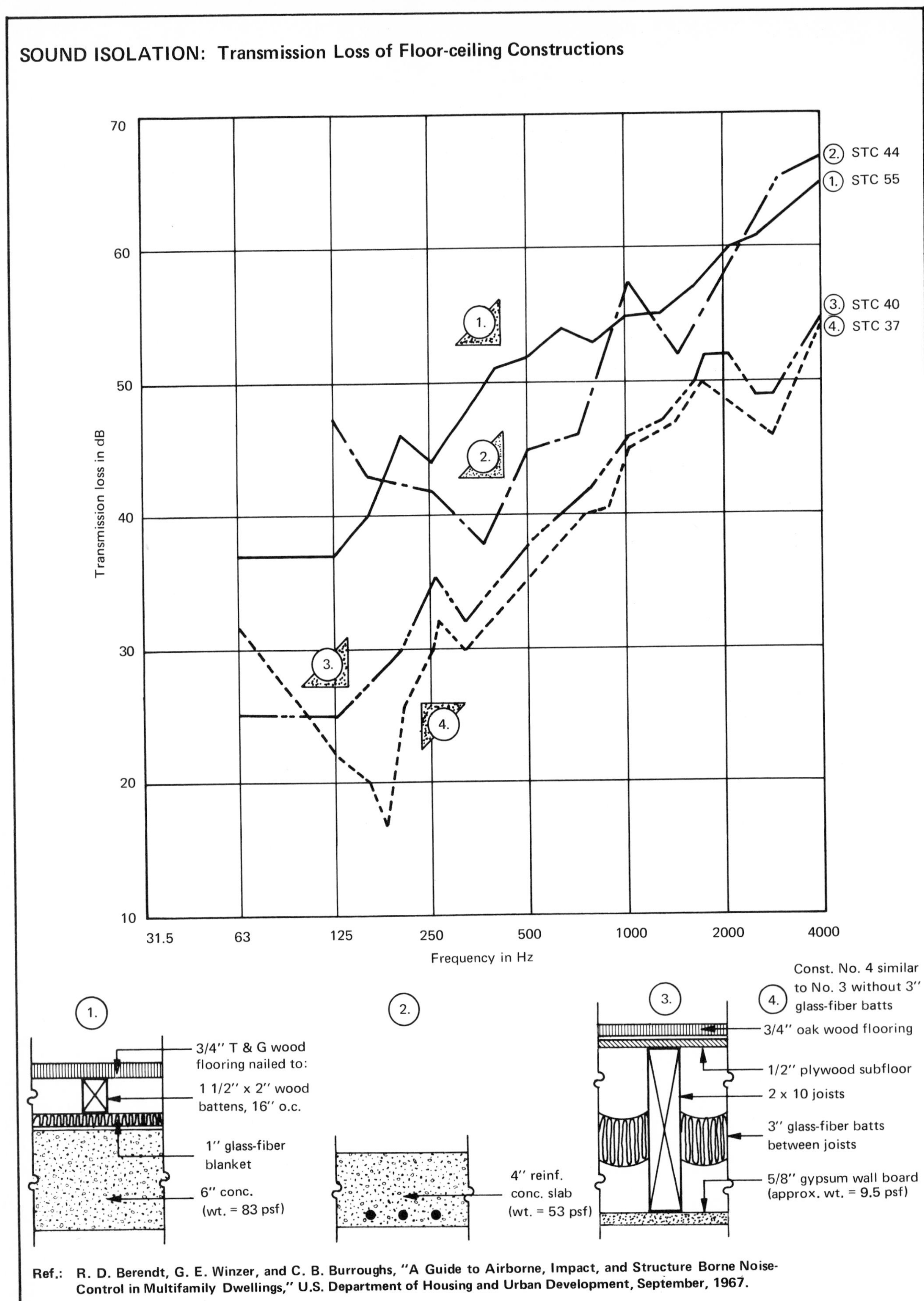

Ref.: R. D. Berendt, G. E. Winzer, and C. B. Burroughs, "A Guide to Airborne, Impact, and Structure Borne Noise-Control in Multifamily Dwellings," U.S. Department of Housing and Urban Development, September, 1967.

SOUND ISOLATION: Ceilings

The room-to-room ceiling sound path must be as good a barrier as the common wall. Lightweight, porous sound-absorbing tiles and panels are relatively poor isolators. However, materials are available that can provide both effective sound absorption and isolation. These are often high-density sound-absorbing materials backed with a laminate or foil layer to decrease porosity. Refer to the AIMA's "Performance Data" Bulletin (published annually) for information on ceiling tests and ceiling STC (also called "ceiling attenuation factor") data.

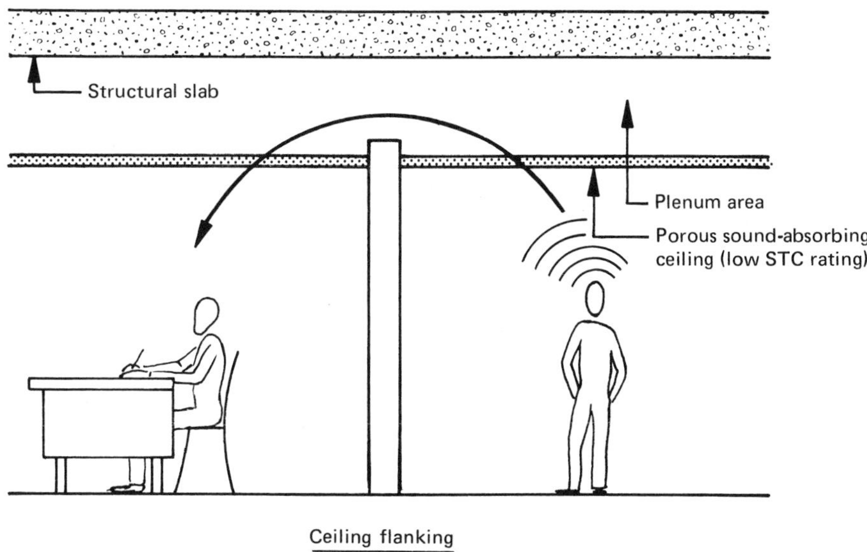

Ceiling flanking

Prevent ceiling flanking by using high STC sound-absorbing materials, or use construction details shown below.

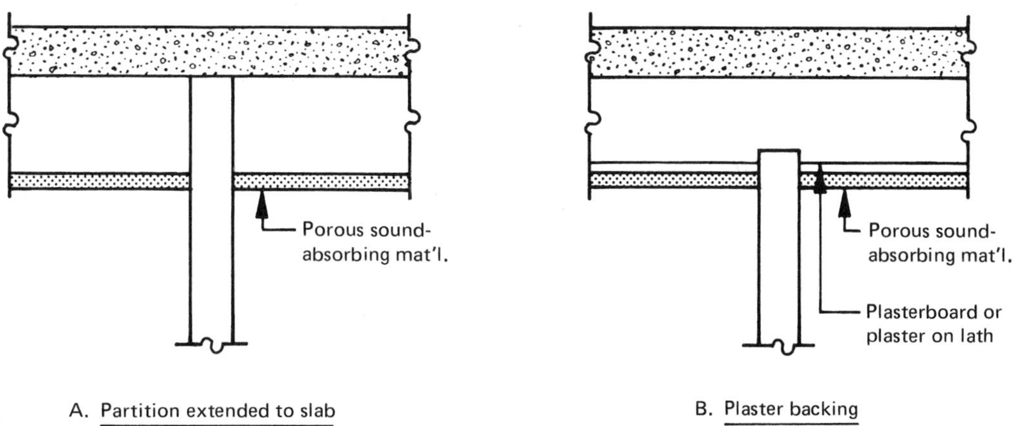

A. Partition extended to slab

B. Plaster backing

NOTE: **Plasterboard or lead sheet plenum barriers that extend from top of partition to structural ceiling slab can also be used to close off plenum flanking path.**

SOUND ISOLATION: Doors

Transmission loss in dB

50
40
30
20
10
0

A.
B.
C.

31.5 63 125 250 500 1000 2000 4000 8000

Frequency in Hz

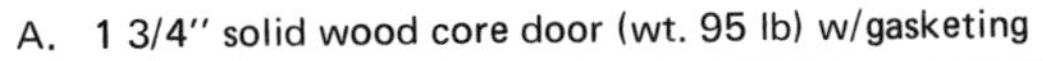

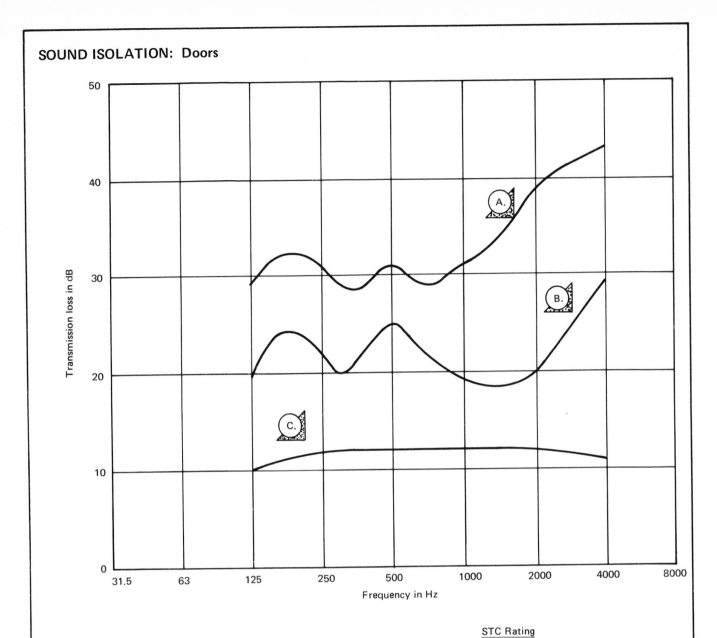

	STC Rating
A. 1 3/4" solid wood core door (wt. 95 lb) w/gasketing	34
B. 1 3/4" hollow wood core door (wt. 30 lb), no gasketing	19
C. Louvered door (25_30% open area)	12

Sound-absorbing ceiling treatment in shaded area

A stub corridor with two doors which provides compatible noise reduction with wall and floor-ceiling sound paths

Sound lock

Gasket all doors, see detail (G)

TV studio

TV studio

Stagger doors across corridors

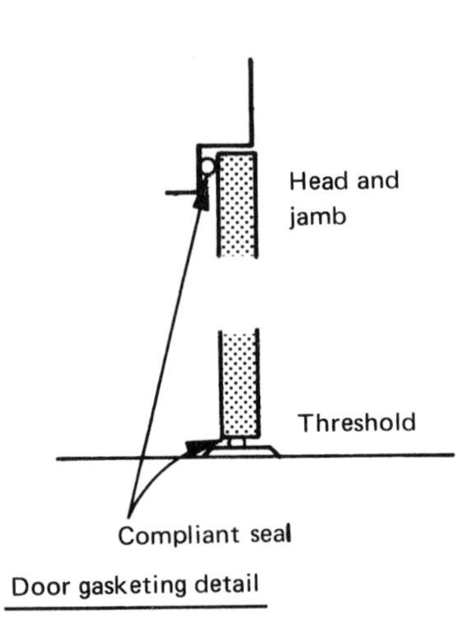

(G) Door gasketing detail

SOUND ISOLATION: Transmission Loss of Glass

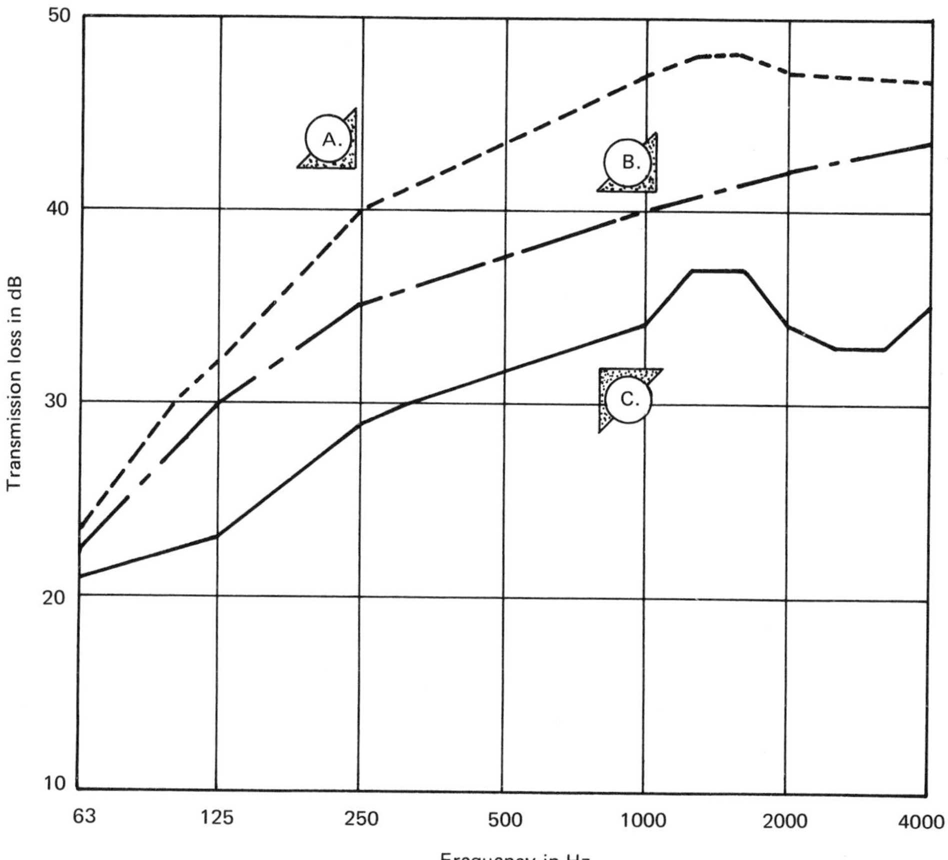

	STC Rating
A. 1/4" + 1/8" double plate-glass window w/4 3/4" air space	43
B. Same as A but w/2" air space	39
C. 1/4" plate-glass window set in caulking	33

To improve TL of windows:

1. Increase thickness of single pane up to 1/4".
2. Use laminated glass.
3. Use double window construction w/air space and panes of different thickness to avoid resonance effects.

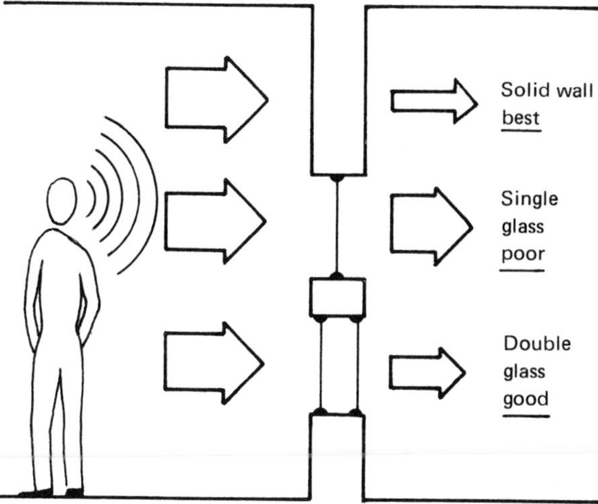

SOUND ISOLATION: Approximate TL Design Curve

The procedure outlined below may be used to estimate the TL of common homogeneous materials. It should only be used if TL data from laboratory tests or reliable field measurements are unavailable.

Procedure:

1. Find TL at 400 Hz by formula TL = 19 + 20 log G, where G equals surface density in psf.
2. Draw 6 dB/octave curve through this point.
3. Draw horizontal line at plateau height of material (see below). From intersection with 6 dB/octave curve, measure plateau breadth.
4. At end of plateau breadth, draw 10 dB/octave curve.

Material	Surface Density, psf/in.	Plateau Height, dB	Plateau Breadth, octaves
Aluminum	14	29	3.7
Brick	11	37	3.3
Cinder Block	6	28	2.7
Dense Concrete	12	38	3.3
Fir Plywood	3	19	2.7
Glass	14	27	3.3
Lead	59	56	2.3
Sand Plaster	9	30	3.0
Steel	40	40	3.7

Ref.: B. G. Watters, TL of Some Masonry Walls, J. Acoust. Soc. Amer., Vol. 31, No. 7, July, 1959.

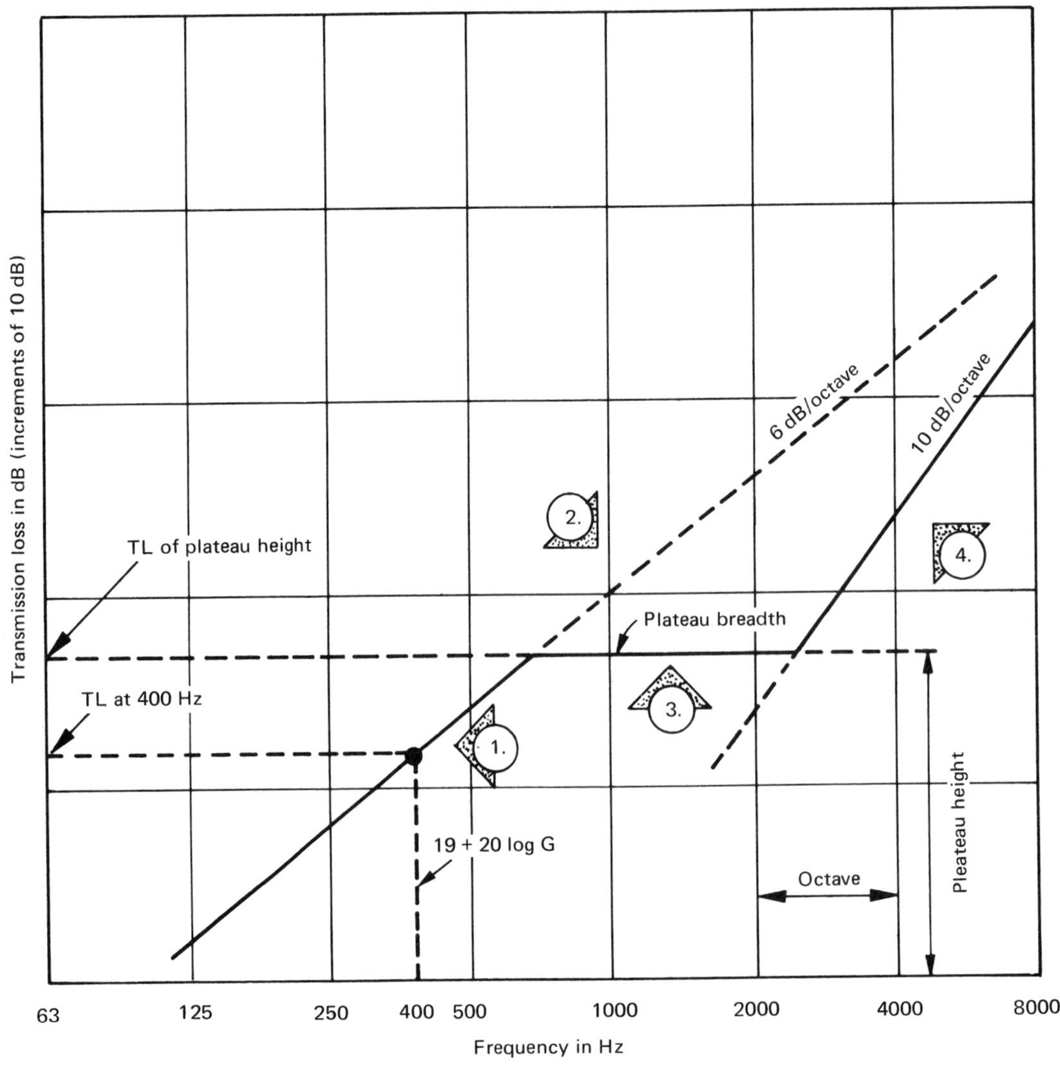

WALL STC IMPROVEMENTS

The following table shows typical wall modifications along with the corresponding estimated improvements in STC ratings.

Modification to Basic Construction	Typical STC Improvement
1. Concrete block with sand-filled cells	3
2. Sound-absorbing material, at least 1-in. thick glass-fiber or mineral wool, in cavity of drywall constructions	5
3. Plaster skin at least 1/2-in. thick or similar surface skin on painted concrete block	2
4. Improvement No. 3, both sides	4
5. Rigidly furred surface skin (for example, 1/4-in. plaster or gypsum board) on concrete block	5
6. Improvement No. 5, both sides	8
7. Resiliently supported (e.g., steel channels or clips) skin (Avoid fastening cabinets, fixtures, etc., which can "short out" the resilient supports!)	10
8. Improvement No. 7, both sides	12
9. Sound-absorbing material in cavity of improvements Nos. 5 and 7	3
10. Sound-absorbing material in cavities of improvements Nos. 6 and 8	5
11. Dividing wall into separate wythes	(See page 77)

Example: Estimate the improvement in the STC of a 6-in. painted concrete-block wall from adding a resiliently supported 1/4-in. plaster skin to one side.

	STC
6-in. painted concrete block	44
Improvement No. 7, above	+10
6-in. painted concrete block wall, with resiliently supported skin	54

NOTE: TL data from field measurements or laboratory tests for the actual construction are *always* preferred.

SOUND ISOLATION: Density Myth (for cavity insulation in walls)

Tests by Owens-Corning Fiberglas Corp. and Riverbank Laboratories show that the density of cavity insulation has little effect on a wall's STC rating. The table at the bottom of the page shows STC results for three wall constructions tested under identical conditions with 3/4 pcf and 2 3/4 pcf (avg.) density glass-fiber cavity insulation.

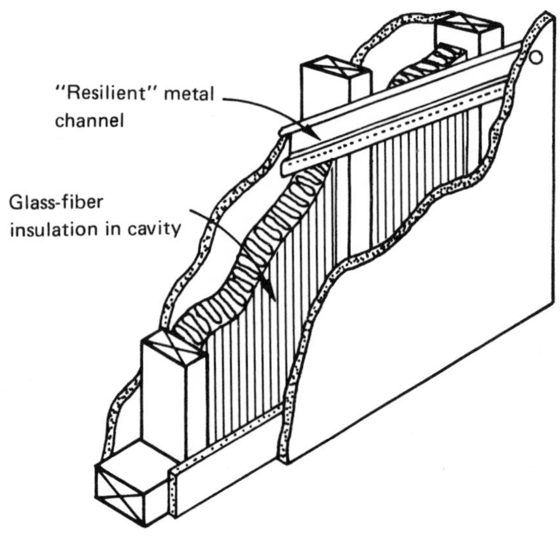

Glass-fiber insulation in cavity

A. 2 x 4 wood studs w/5/8" layer gyp. board on both sides

B. 2 1/2" steel channel studs w/2 layers 1/2" gyp. board on one side and one layer on other side

C. Same as B w/3 5/8" steel channel studs

	STC Ratings		
Construction	**Without Insulation**	**3/4 pcf Density Glass Fiber**	**2 3/4 Density Glass Fiber**
A	44	50	50
B	44	51	51
C	45	49	49

Ref.: "Solutions to Noise Control Problems," Owens-Corning Fiberglas Corp., Toledo, Ohio, 1969.

SOUND ISOLATION: Noise Reduction and Masking Sound(Background Noise)

Background noise in rooms can contribute to practical sound isolation design by "masking" intruding noise. For example, if a source room (1.) noise level (SPL_1) of 80 dB must be reduced below a background noise level (NC) of 20 dB in receiving room (2.) by about 10 dB, the required NR between rooms will be 80–(20–10)=70 dB (See heavy solid line on figure). However, by raising the receiving room background noise level to 35 dB the required NR will now be 80–(35–10)=55 dB. This is a significant NR (and STC rating) reduction of 70–55=15 dB! The following page introduces the concept of noise criteria which is a method for establishing acceptable background noise levels in rooms.

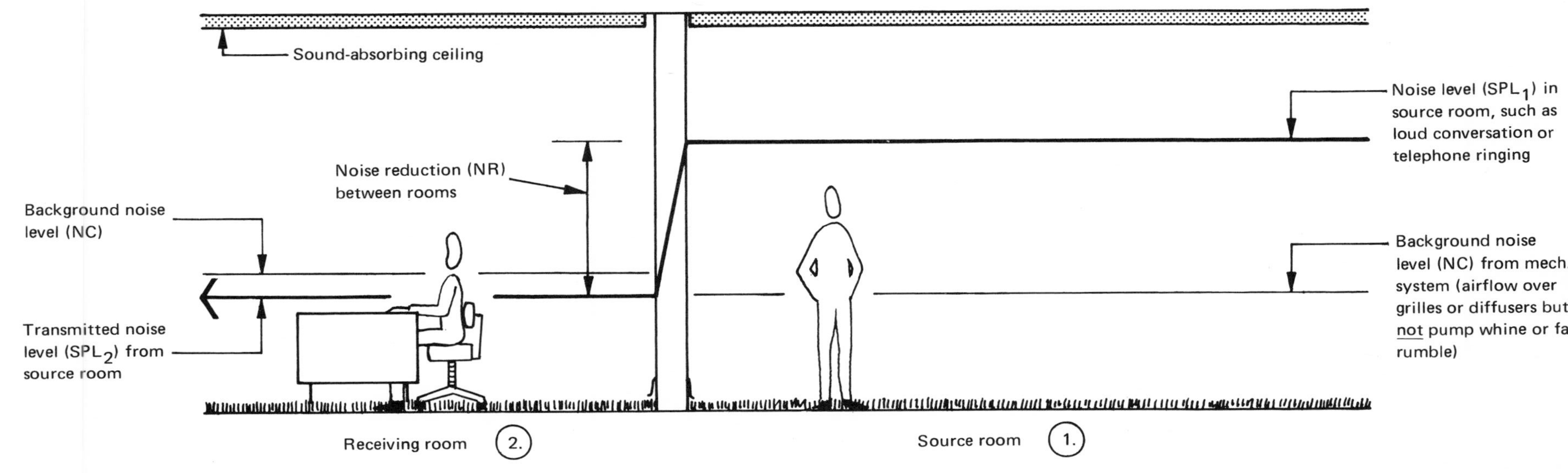

- **If the transmitted noise level is under the background noise level in the receiving room, as shown above, it will be masked or "covered up" by the normal existing background noise level (called "ambient" sound). Masking sound should be bland and continuous so people will scarcely notice it.**

- **For good isolation, receiving room NC $\geqslant$ SPL_2 ($\geqslant$ means greater than or equal to), where NC represents desired background SPL at a given frequency.**

RECOMMENDED NOISE CRITERIA FOR ROOMS

Noise criteria (NC) curves can be used as a method for specifying continuous background noise levels to achieve sound isolation (and for evaluating existing noise situations as well). The table below presents recommended NC ranges for various indoor functional activity areas. Each NC curve is defined by its sound pressure level for the eight octave-band frequencies shown by the graph on the following page. Note that the NC rating for a noise situation usually means the lowest NC curve that is not exceeded by any octave-band sound pressure level.

Type of Space (and listening requirements)	Preferred Range of Noise Criteria*	Equivalent dBA Level†
Concert halls, opera houses, recording studios, recital halls, etc. (for excellent listening conditions)	NC-15 to NC-20	25 to 30
Bedrooms, sleeping quarters, hospitals, residences, apartments, hotels, motels, etc. (for sleeping, resting, relaxing)	NC-20 to NC-30	30 to 40
Auditoriums, theaters, radio/TV studios, music practice rooms, large meeting rooms, audio-visual facilities, large conference rooms, executive offices, churches, courtrooms, chapels, etc. (for very good listening conditions)	NC-20 to NC-30	30 to 40
Private or semiprivate offices, small conference rooms, classrooms, reading rooms, libraries, etc. (for good listening conditions)	NC-30 to NC-35	40 to 45
Large offices, reception areas, retail shops and stores, cafeterias, restaurants, gymnasiums, etc. (for fair listening conditions)	NC-35 to NC-40	45 to 50
Lobbies, corridors, laboratory work spaces, drafting and engineering rooms, general secretarial areas, maintenance shops such as for electrical equipment, etc. (for moderately fair listening conditions)	NC-40 to NC-45	50 to 55
Kitchens, laundries, school and industrial shops, garages, machinery spaces, computer equipment rooms, etc.	NC-45 to NC-55	55 to 65

*It is good practice to design the ambient acoustical background with a minimum-maximum NC range as suggested in the table. The *minimum* background noise levels are important to achieve in order to contribute to the sound isolation desired. For example, by controlling the pressure drop available at the mechanical system room air supply registers or grilles, one can often generate just the right amount of unobtrusive background noise. The *maximum* background noise levels should not be exceeded. This latter restriction will prevent the background itself from becoming a source of annoyance.

†Do not use dBA values for specification purposes. For example, identical A-scale readings can be obtained for backgrounds having a wide variety of spectrum shapes.

SOUND ISOLATION: Noise Criteria (NC) Curves

The rating of an NC level for a noise is found by comparing its sound spectrum with the NC curves shown below. Its SPL's may exceed an NC curve by up to +3 dB in one or two bands if in two to four other bands the level is –3 or more dB below the NC curve. The NC rating then is the highest NC curve meeting these requirements.

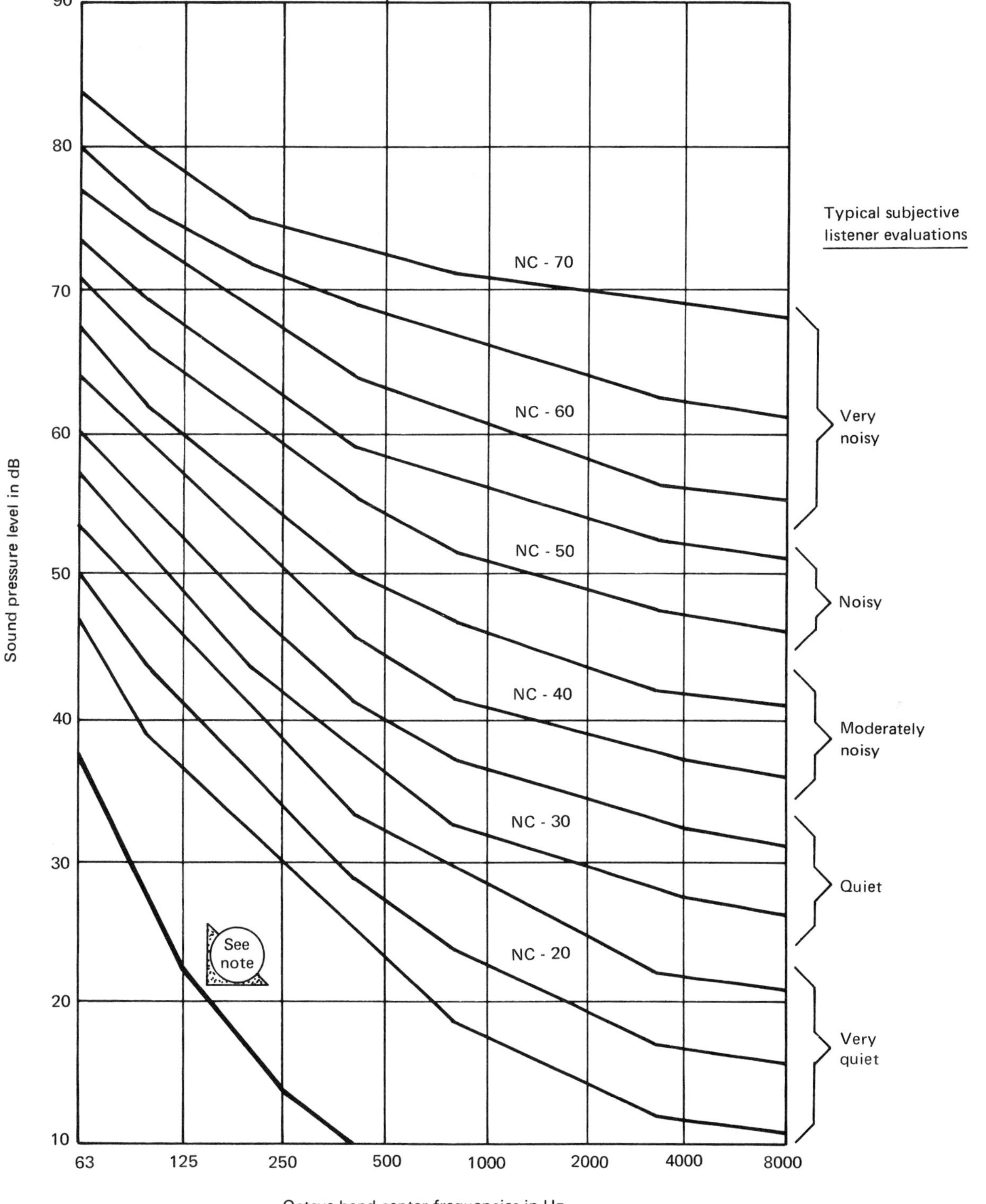

NOTE: **Dark curve indicates threshold of hearing for continuous noise (Ref.: Acustica, Vol. 14, p. 33, Fig. 14, 1964). Threshold and NC curves have low frequency sound levels greater than high frequency levels because of lower human perception of low frequency sound and corresponding greater tolerance.**

NOISE CRITERIA SOUND PRESSURE LEVEL TABLE*

NC Curve	Sound Pressure Level, dB							
	63 Hz	125 Hz	250 Hz	500 Hz	1000 Hz	2000 Hz	4000 Hz	8000 Hz
NC-70	83	79	75	72	71	70	69	68
NC-65	80	75	71	68	66	64	63	62
NC-60	77	71	67	63	61	59	58	57
NC-55	74	67	62	58	56	54	53	52
NC-50	71	64	58	54	51	49	48	47
NC-45	67	60	54	49	46	44	43	42
NC-40	64	57	50	45	41	39	38	37
NC-35	60	52	45	40	36	34	33	32
NC-30	57	48	41	36	31	29	28	27
NC-25	54	44	37	31	27	24	22	21
NC-20	50	41	33	26	22	19	17	16
NC-15	47	36	29	22	17	14	12	11

*For convenience in using noise criteria data, the table lists the sound pressure levels (SPL's) in decibels for the NC curves from the preceding page.

SOUND ISOLATION: Example Problem—Transmission Loss (TL) Design

Given:

10′ x 10′ x 10′ reading room, "fuzz" ceiling and carpeted flr., painted concrete walls with computer room adjacent. (See sketch below.)

- Find absorption in receiving room, i.e., reading room, using formula a (sabins) = surface area X coefficient. (See table below.)

Area	Octave-band Center Frequencies, Hz 125	250	500	1000	2000	4000
100 sq ft ceiling, 1 1/2" fuzz	38	60	78	80	78	70
100 sq ft floor, carpet on foam rubber	8	24	57	69	71	73
400 sq ft walls, painted conc. blk.	40	20	24	28	36	32
a_2, sabins:	86	104	159	177	185	175

- To find req. NR, subtract background level in receiving room from likely noise in source room. Req. TL can be found from formula TL = NR − 10 log a_2/S at each frequency. Then select wall construction with TLs greater than req. TLs.

	125	250	500	1000	2000	4000	See Page
Likely noise (SPL_1):							
Likely noise in computer room	75	73	78	80	78	74	21
Background level:							
Min. NC-30 in reading room	48	41	36	31	29	28	88
Req. NR, dB:	27	32	42	49	49	46	
Less 10 log a_2/S:	−1	0	2	3	3	3	67
Req. TL, dB:	28	32	40	46	46	43	
Use 6" conc. blk. wall, painted. Wt. = 34 psf — TL (dB):	37	36	42	49	55	58	73

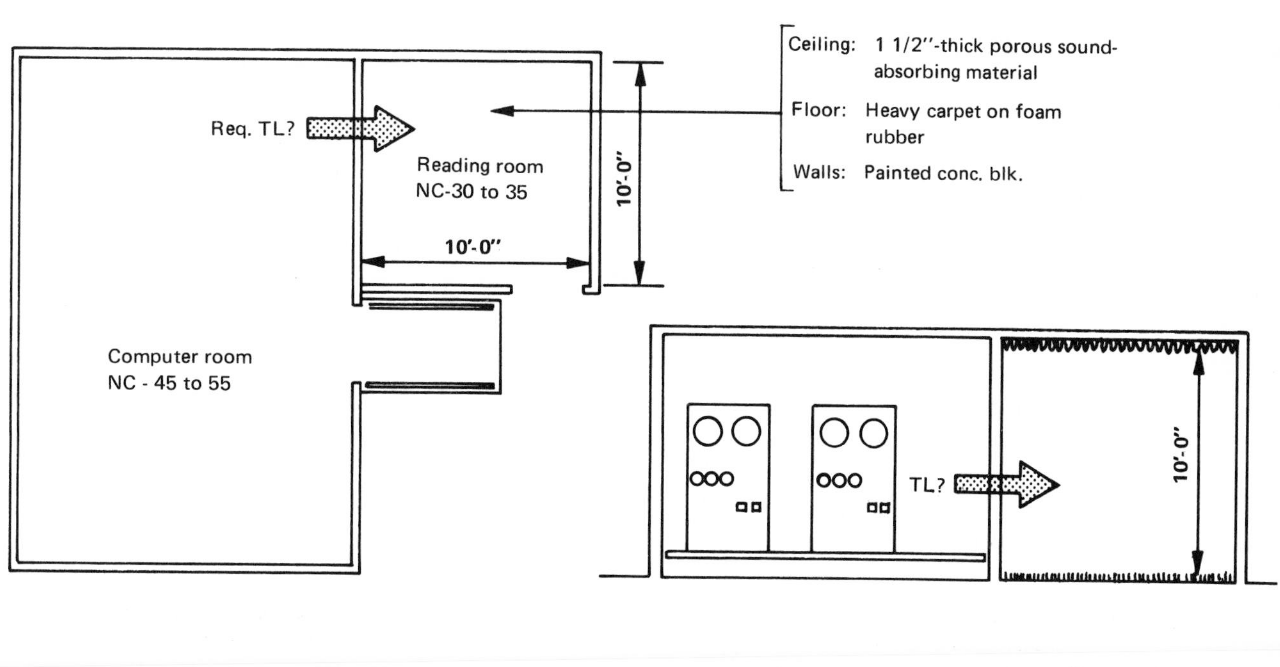

SOUND ISOLATION: Suggested Sound Transmission Class (STC) Ratings for Partitions

Suggested STC's can be found in the table below at the intersection of the desired source room row and receiving room column. For example, an STC 65 would normally be required between a kitchen and a classroom. Be careful; a receiving room can also be a source room, so check both paths!

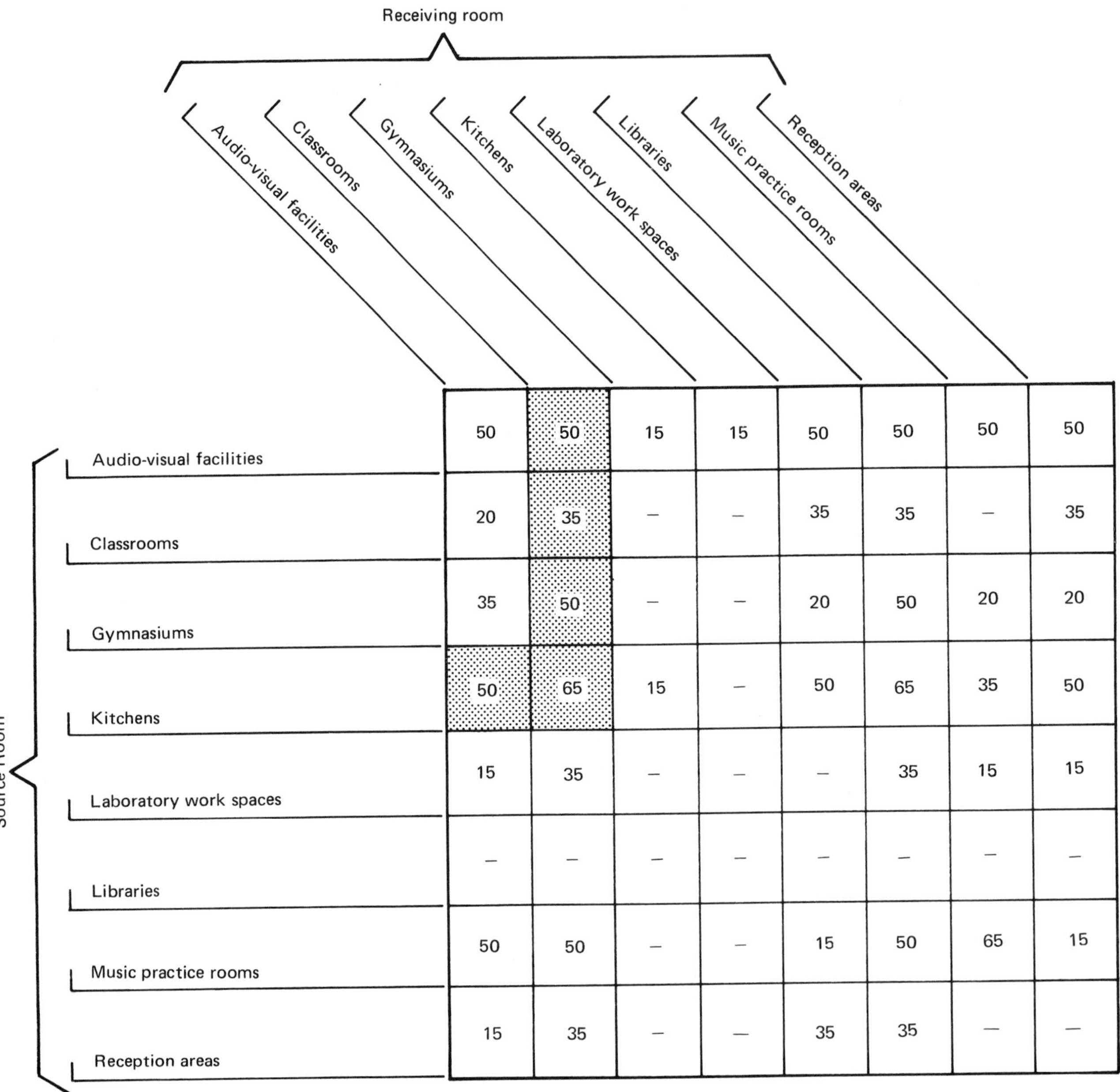

Source Room \ Receiving room	Audio-visual facilities	Classrooms	Gymnasiums	Kitchens	Laboratory work spaces	Libraries	Music practice rooms	Reception areas
Audio-visual facilities	50	50	15	15	50	50	50	50
Classrooms	20	35	–	–	35	35	–	35
Gymnasiums	35	50	–	–	20	50	20	20
Kitchens	50	65	15	–	50	65	35	50
Laboratory work spaces	15	35	–	–	–	35	15	15
Libraries	–	–	–	–	–	–	–	–
Music practice rooms	50	50	–	–	15	50	65	15
Reception areas	15	35	–	–	35	35	–	–

NOTE: **This table indicates an approximation of the required sound isolation between listed activities on a STC basis. It assumes that the receiving room actually attains a minimum ambient background level according to the recommended noise criteria in this section.**

SOUND ISOLATION: Barriers Between Exterior Sound Source and Interior Receiver

Barriers can be effectively used to reduce outdoor noise, particularly high frequency sound such as vehicle tire whine. However, low frequency sound such as vehicle rumble and engine roar tends to diffract past barriers. Actual attenuation in decibels can be estimated for various barrier configurations by the curves on the following page.

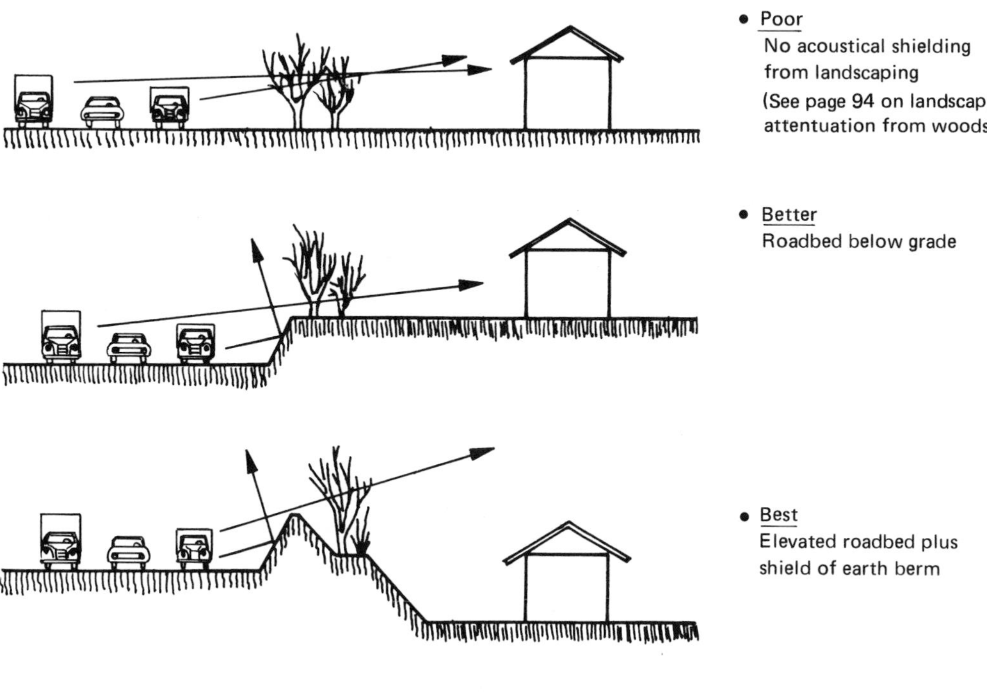

Natural barriers

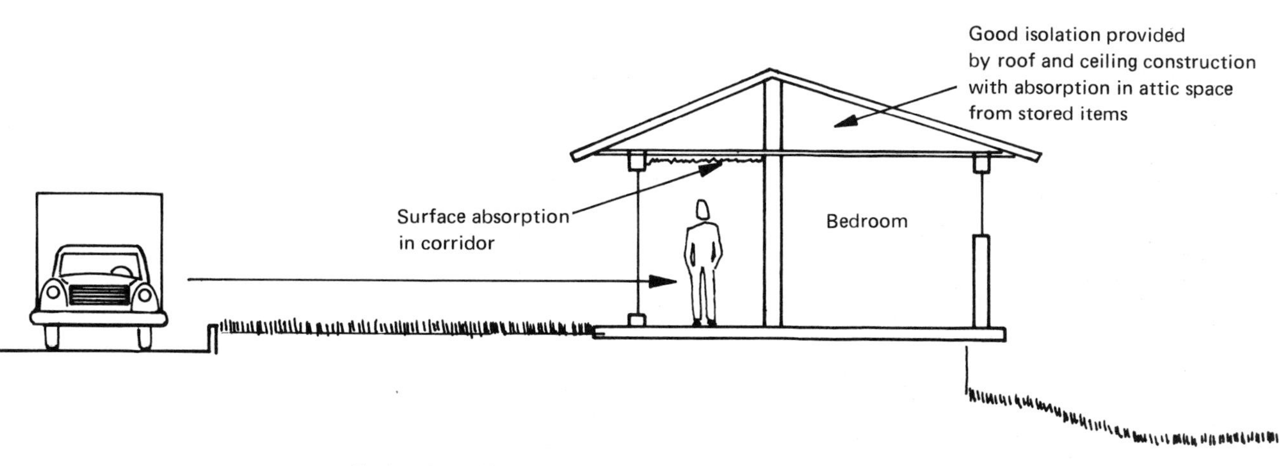

Enclosed corridor as acoustical barrier

NOTE: **Heavy vehicular traffic represents a continuous line source which radiates sound cylindrically (not spherically as from a point source). Consequently, car noise decreases by only 3 dB for each doubling of distance from the source. Trucks, however, can usually be treated as point sources with 6 dB decay for each doubling of distance from the inverse-square law.**

SOUND ISOLATION: Attenuation Outdoors Due to a Barrier

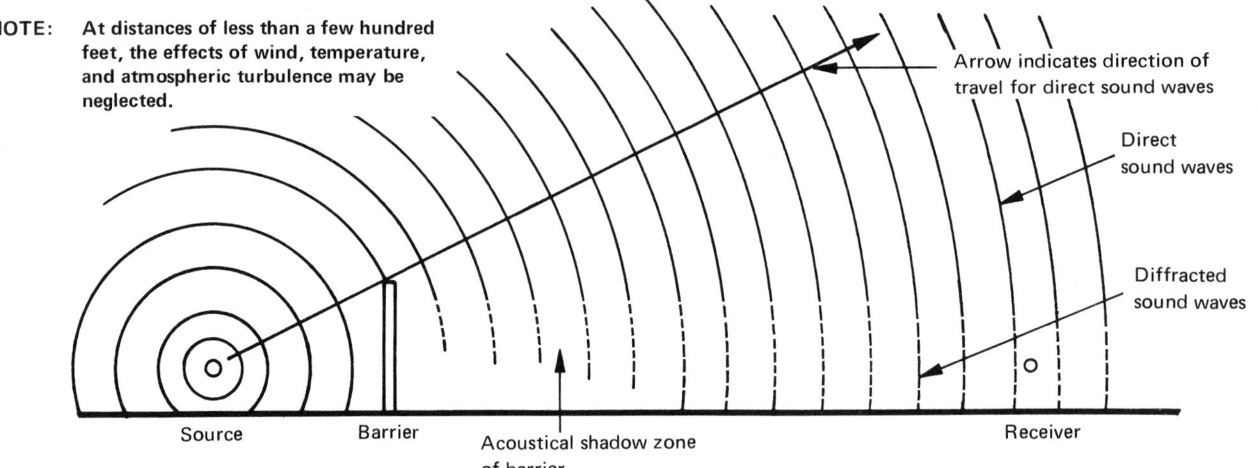

To effectively use barriers:

1. Place barrier as close as possible to either sound source or receiver.
2. The greater the height of the barrier above the direct source-receiver path, the greater the attenuation (i.e., reduction of sound levels).
3. Barrier should be solid and airtight. However, the barrier material's TL need not exceed attenuation from the barrier configuration.
4. Attenuation shown on curves below is for barriers of infinite length. Sound diffraction around ends of short barriers will reduce effectiveness considerably. Provide total length at least 4R.

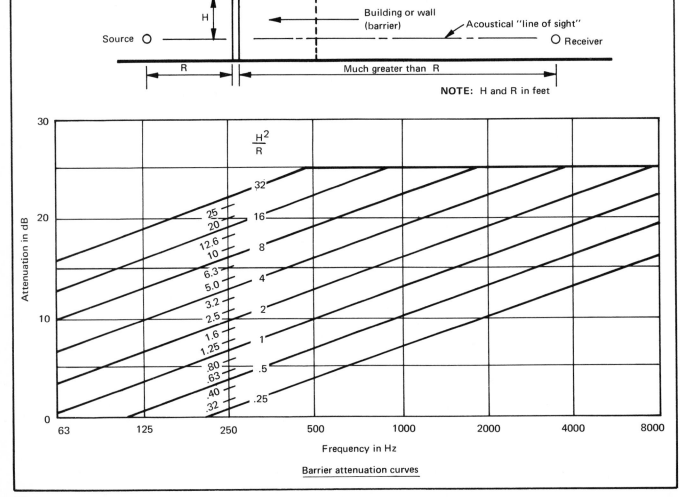

H

Building or wall (barrier)

Acoustical "line of sight"

Source

Receiver

R

Much greater than R

NOTE: H and R in feet

$\frac{H^2}{R}$: 32, 25, 20, 16, 12.6, 10, 8, 6.3, 5.0, 4, 3.2, 2.5, 2, 1.6, 1.25, 1, .80, .63, .5, .40, .32, .25

Attenuation in dB: 0, 10, 20, 30

Frequency in Hz: 63, 125, 250, 500, 1000, 2000, 4000, 8000

Barrier attenuation curves

SOUND ISOLATION: Use of Attenuation Curves for Outside Barriers

H \ R	1	2	3	4	5	6	7	8	9	10
0.5	0.25	—	—	—	—	—	—	—	—	—
1	1	0.5	0.3	0.3	0.2	0.2	—	—	—	—
1.5	2.3	1.1	0.8	0.6	0.5	0.4	0.3	0.3	0.3	0.2
2	4	2	1.3	1.0	0.8	0.7	0.6	0.5	0.4	0.4
2.5	6.3	3.1	2.1	1.6	1.3	1.1	0.9	0.8	0.7	0.6
3	9	4.5	3	2.2	1.8	1.5	1.3	1.1	1.0	0.9
3.5	12.3	6.2	4.1	3.1	2.5	2.0	1.8	1.5	1.4	1.2
4	16	8	5.3	4	3.2	2.7	2.3	2	1.8	1.6
4.5	20.2	10.1	6.8	5.1	4	3.4	2.9	2.5	2.3	2.0
5	25	12.5	8.3	6.3	5	4.2	3.6	3.1	2.8	2.5

Values of H^2/R, where H and R are in feet

Example — use of barrier curves:

Given: **Source of noise 3′-0″ from barrier, which extends 4′-0″ above source.**
$H^2/R = (4)^2/3 = 5.3$ (see chart above)

Sound attenuation in dB

Frequency (Hz):	63	125	250	500	1000	2000	4000	8000
Attenuation in dB: (from curves on preceding page)	8	11	14	17	20	23	25	25

SOUND ISOLATION: Effectiveness of Landscaping as Sound Barrier

Trees and vegetation are not normally effective as sound barriers. For example, dense planting at least 100 feet deep will provide only 7 to 11 dBs of sound attenuation as shown by the curve below. Also, deciduous trees provide almost no shielding during the months when their leaves have fallen.

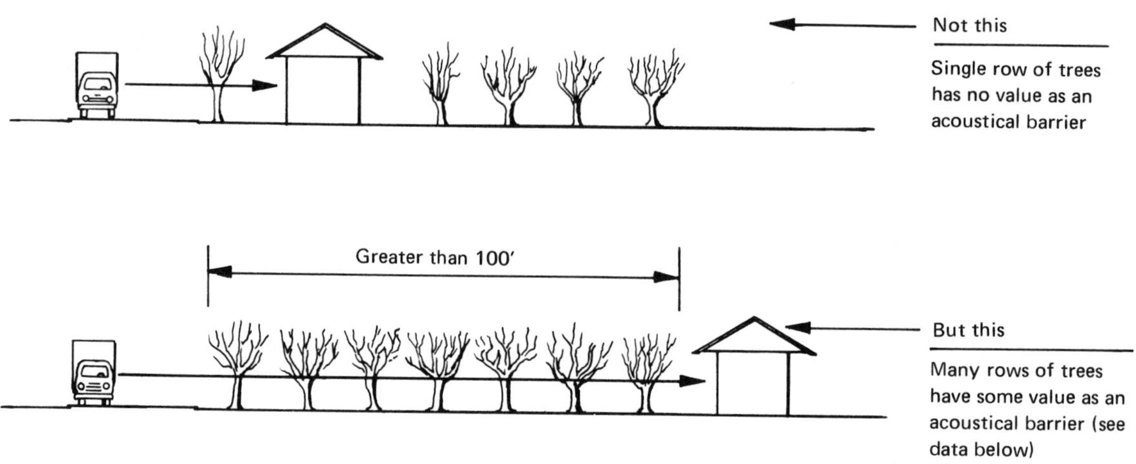

Ref.: T. F. W. Embleton, Sound Propagation in Homogeneous Deciduous and Evergreen Woods, J. Acoust. Soc. Amer., Vol. 35, No. 8, August, 1963.

Sound attenuation in dB per 100 ft

20
15
10
5
0

63 125 250 500 1000 2000 4000 8000

Frequency in Hz

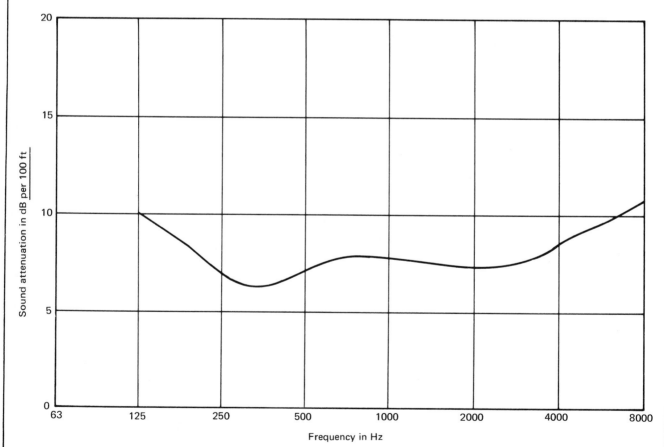

Typical noise attenuation of deciduous and evergreen woods per 100 ft

NOTE: **It was also found by Embleton that amplification (no mid-frequency attenuation) occurred within 50 ft of the sound source due to resonance of tree trunks and branches!**

SOUND ISOLATION: Attenuation Outdoors (Temperature and Wind)

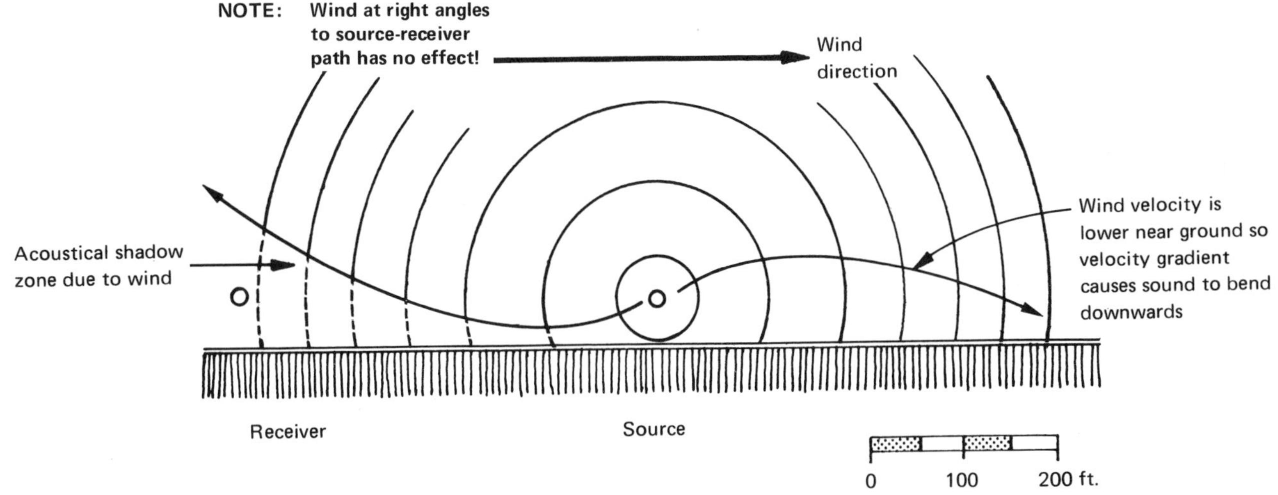

WIND

The effect of wind on sound outdoors is a complex phenomenon. Downwind from the source, sound is normally bent toward the ground increasing its level. Upwind, sound is bent upwards causing a shadow zone where its level is reduced. For example, at distances of about 500 ft. (as shown above) the upwind mid-frequency attenuation for winds of 10 mph can be about 10 dB. Note that a reversal of wind direction can increase the level by about 10 dB at the same location! Do not rely on wind attenuation in design.

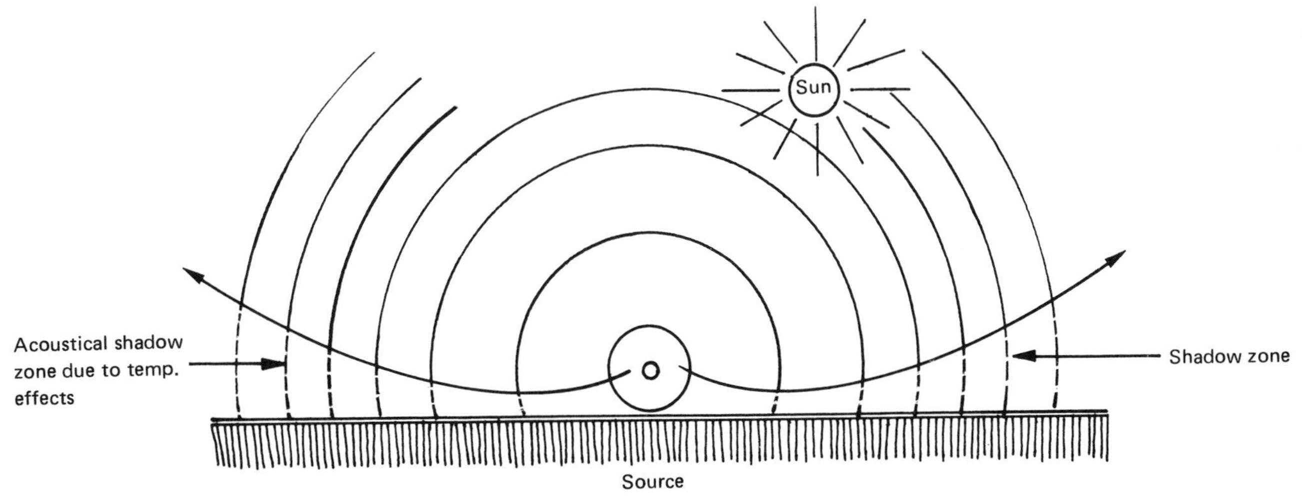

TEMPERATURE

On a clear, calm day the effect of temperature gradients (i.e., decrease of air temp. with elevation) can cause sound to bend upwards as shown by the above sketch. Conversely, on a clear, calm night sound will tend to bend toward the ground.

NOTE: Estimates of attenuation outdoors due to temperature and wind depend on meterological as well as acoustical data. For further information on sound propagation outdoors, see p. 164–193, L. L. Beranek, Noise and Vibration Control, McGraw-Hill, New York, 1971.

SUGGESTED GENERAL CLASSIFICATION FOR ROOMS

*Type of Space**	*Noisy Source†*	*Quiet Ambient Background‡*
Auditoriums, theaters, radio/TV studios, music practice rooms, large meeting rooms, classrooms, audio-visual facilities, conference rooms, etc.	Yes	Yes
Residences, apartments, hotels, motels, reception areas, churches, courtrooms, chapels, large offices, etc.	Sometimes	Yes
Private offices, reading rooms, libraries, bedrooms, sleeping quarters, hospitals, etc.	No	Yes
Retail shops and stores, cafeterias, corridors, maintenance shops, kitchens, laundries, school and industrial shops, garages, computer equipment rooms, etc.	Yes	No

*It is good practice to locate the spaces requiring quiet ambient backgrounds (i.e., low NC ranges) as far as possible from noisy spaces or to provide "buffer" spaces between. Try to group together the noisy rooms which do not have quiet background requirements.

†Noise levels for several of the listed spaces are presented on pages 21 & 22.

‡Recommended noise criteria for rooms are given on page 86.

SOUND ISOLATION: Site Orientation

Sometimes buildings can be arranged so that less critical spaces can act as a barrier to exterior noises. The school laboratory work spaces shown below can be used to shield the quieter library space from the noisy urban highway. Also, the classrooms and the office spaces can be used to shield the auditorium from exterior as well as gymnasium noises.

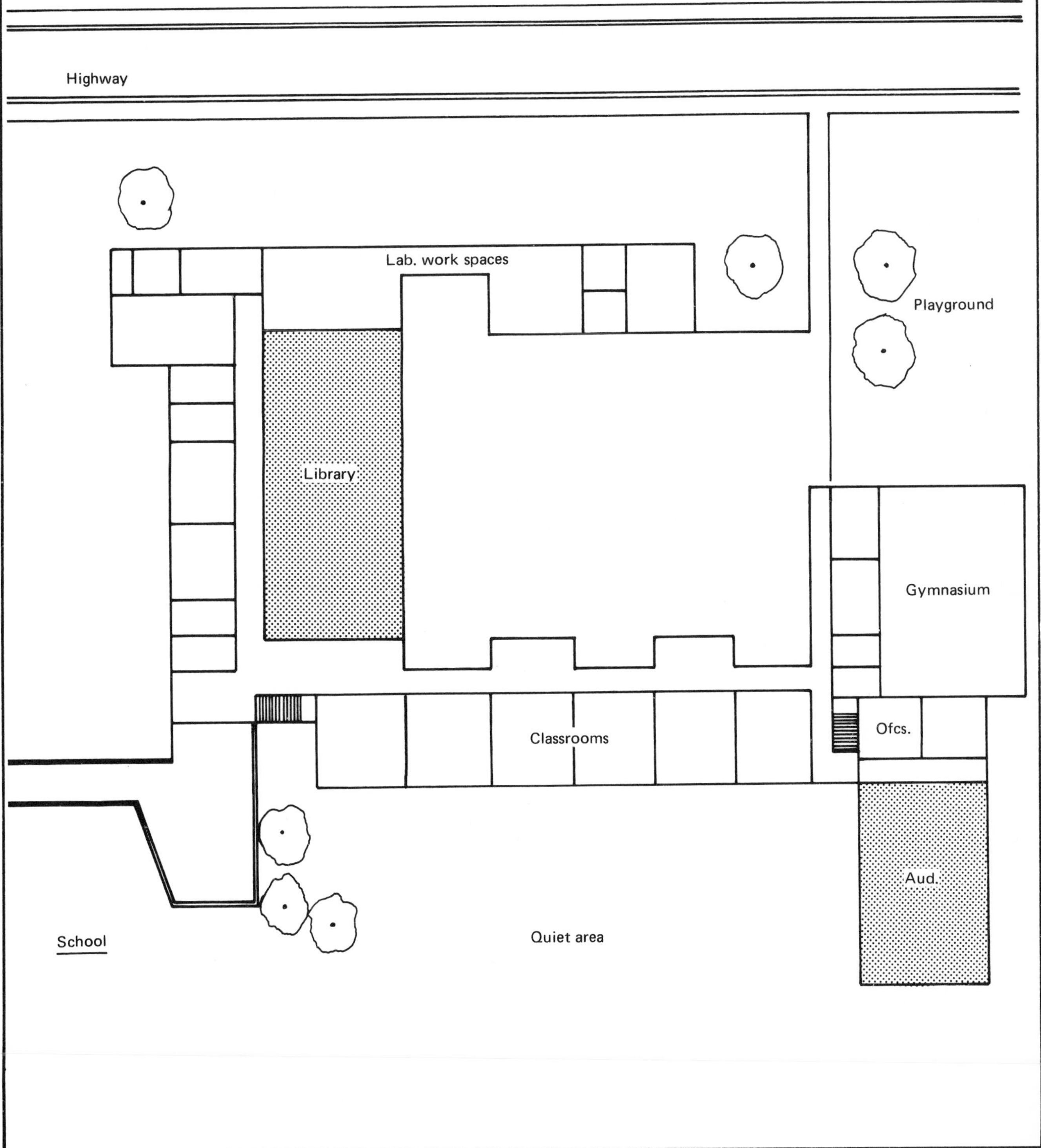

SOUND ISOLATION: Building Orientation

Courtyards can be a source of considerable noise. The apartment buildings shown below have a central courtyard surrounded by parallel walls. These hard-surfaced parallel walls will cause annoying flutter echoes which can intensify the courtyard noise.

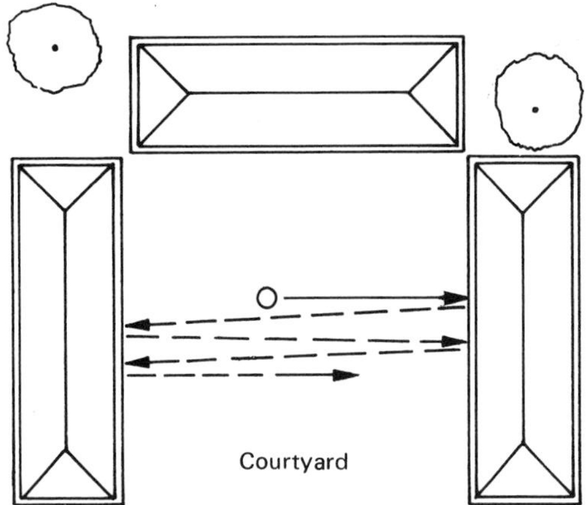

• Poor

Parallel walls of courtyard make exterior noise condition worse by containing noise and causing flutter echoes

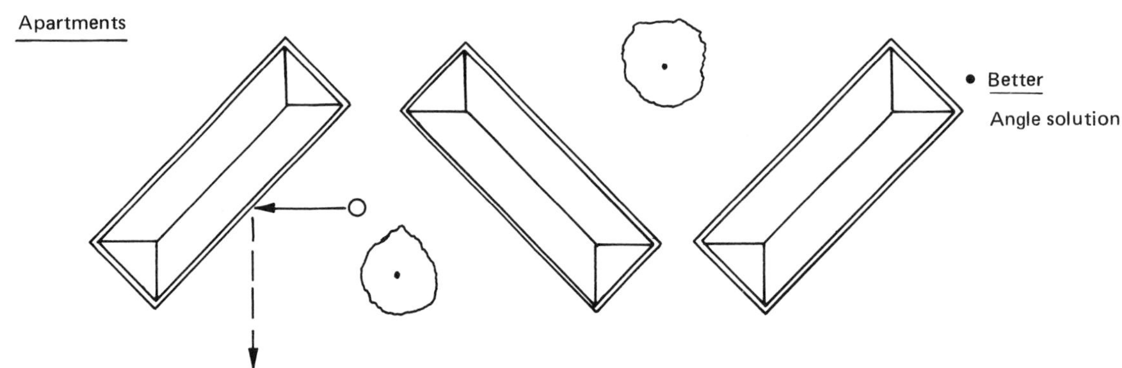

• Better

Angle solution

• Best

Staggered solution

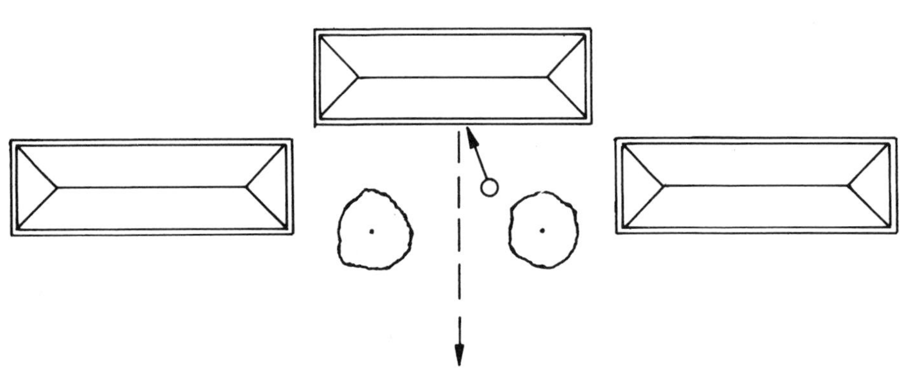

SOUND ISOLATION: Site Location of Buildings (Motels Near an Airport)

In evaluating aircraft noise a quantity called "perceived noise level" in decibels (abbrv. PNdB) has been developed to account for greater human sensitivity to high frequency sound. PNdB's calculated from measured noise levels are more high frequency weighted than dBA's and correlate very well with an average listener's response to aircraft noises of widely varying character. For example, two noises at the same PNdB level, as shown by the table at the bottom of the page, would be judged equally noisy by an average listener, although their sound spectra might be quite different.

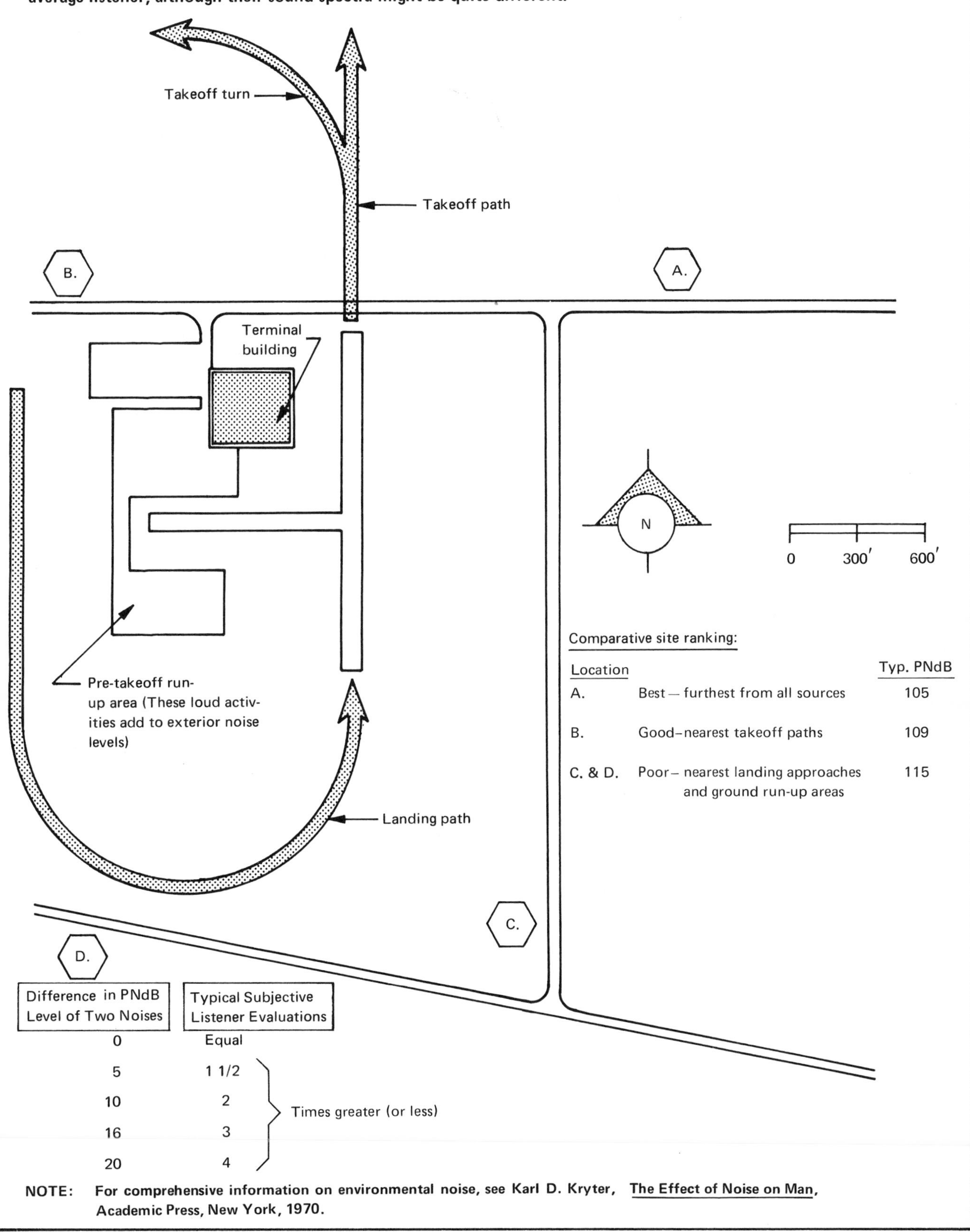

Comparative site ranking:

Location		Typ. PNdB
A.	Best — furthest from all sources	105
B.	Good–nearest takeoff paths	109
C. & D.	Poor– nearest landing approaches and ground run-up areas	115

Difference in PNdB Level of Two Noises	Typical Subjective Listener Evaluations	
0	Equal	
5	1 1/2	Times greater (or less)
10	2	
16	3	
20	4	

NOTE: For comprehensive information on environmental noise, see Karl D. Kryter, The Effect of Noise on Man, Academic Press, New York, 1970.

CHECKLIST FOR EFFECTIVE ISOLATION OF SOUND

- Sound-absorbing materials by themselves normally provide little sound isolation. For example, sound can travel up and over a partition attached to a suspended porous sound-absorbing ceiling, using the open plenum to travel from one room to another.

- Sound will travel through any opening – however small. For effective isolation, provisions should be made to seal cracks or openings in any construction.

- Wall and ceiling constructions should be balanced to provide approximately the same amount of sound transmission loss through each construction.

- Mass is always a major consideration in sound isolation. For example,

  1. Sand is more effective than lightweight aggregates in base-coat plasters.
  2. Dense concrete blocks are more effective than units made with lightweight aggregates. Check density closely, as it may vary widely between different fabrication locations.
  3. The integrity of the mass should never be violated by placing electrical outlets back-to-back or by recessed medicine cabinets, ceiling fixtures, or furniture built into a partition. Avoid back-to-back electrical outlets by staggering them, or pack the outlet box with mineral wool and seal the box airtight to the partition.

- While sound-absorbing materials by themselves are not satisfactory for sound isolation, sound absorption does perform a function of contributing to sound isolation. This is particularly evident at the lower sound frequencies within small rooms where sound originates, such as in music practice rooms.

SECTION 4

Speech Privacy

SOUND ATTENUATION IN SPEECH PRIVACY

IMPORTANT FACTORS

In enclosed rooms:

1. Source room absorption
2. Source room speech use or speech effort
3. Privacy required
4. Receiving room background noise level
5. Noise reduction of construction between rooms

In open plans:

1. Room absorption
2. Speech effort and speaker orientation
3. Privacy required (communication requirements also important in open classrooms)
4. Background noise level
5. Attenuation (or noise reduction) of barriers
6. Distance from source to listener

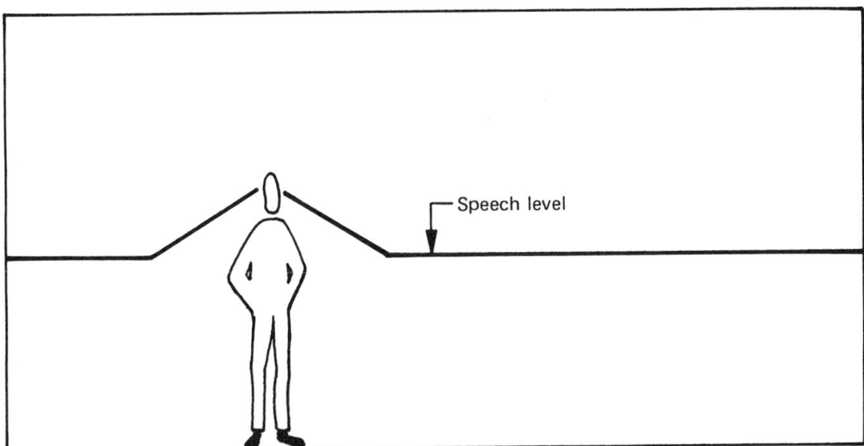

Sound attenuation in closed room w/hard surfaces

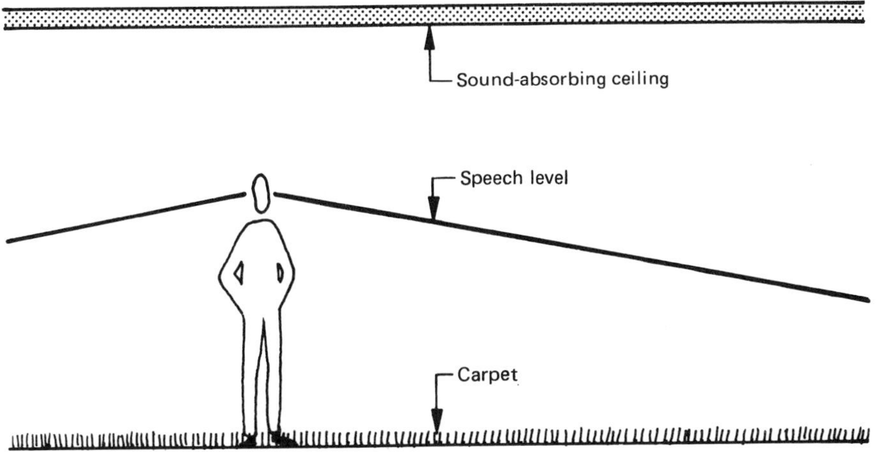

Sound attenuation in open room w/absorptive surfaces*

*In large open plan spaces, where the walls are a relatively insignificant part of the space, the sound attenuation with distance can be 6 dB per doubling of distance with carpeted floors and sound-absorbing ceilings.

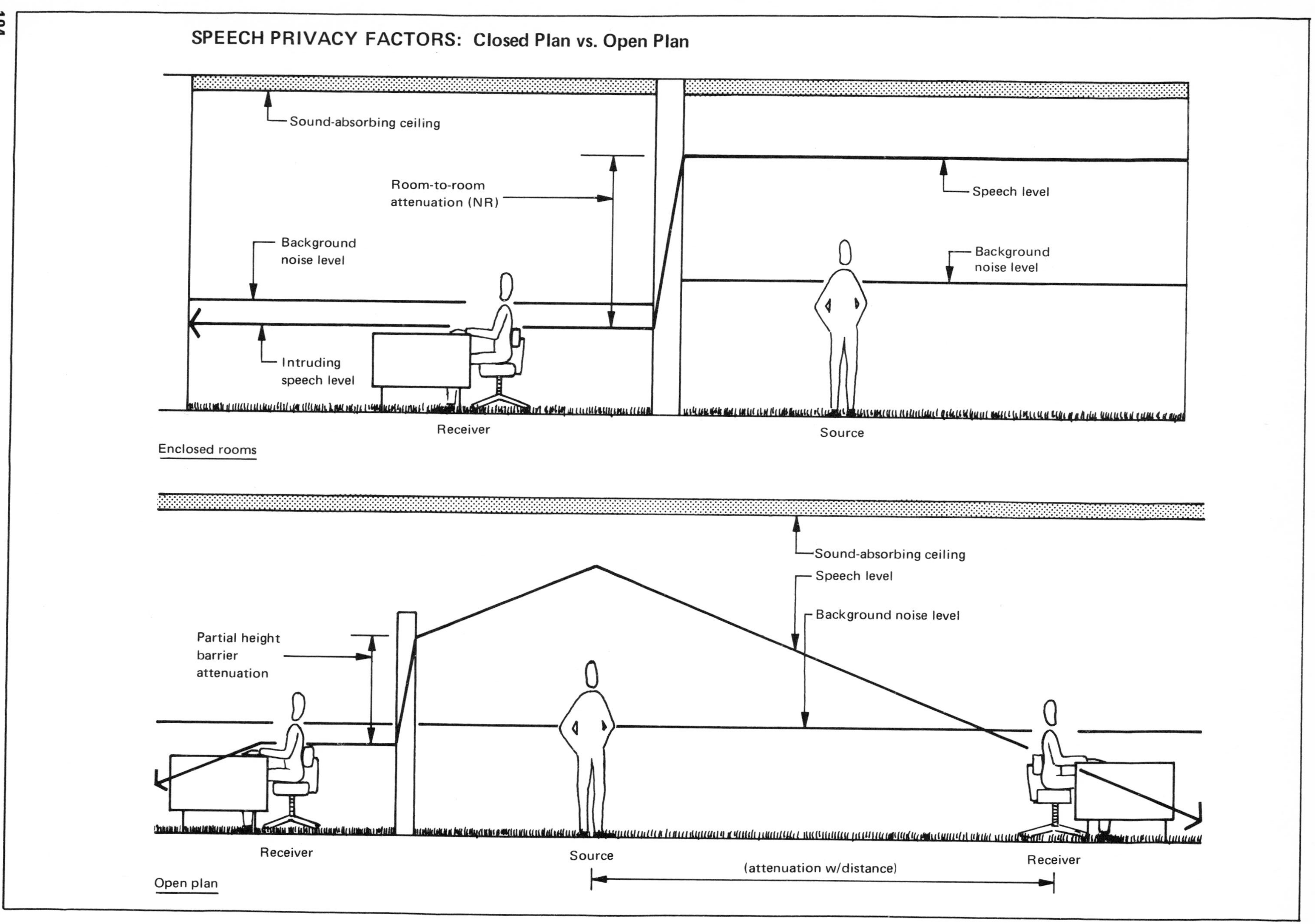
SPEECH PRIVACY FACTORS: Closed Plan vs. Open Plan
Sound-absorbing ceiling
Room-to-room attenuation (NR)
Background noise level
Intruding speech level
Speech level
Background noise level
Receiver
Source
Enclosed rooms
Sound-absorbing ceiling
Speech level
Background noise level
Partial height barrier attenuation
Receiver
Source
(attenuation w/distance)
Receiver
Open plan

SPEECH PRIVACY: Speaker and Listener Orientations

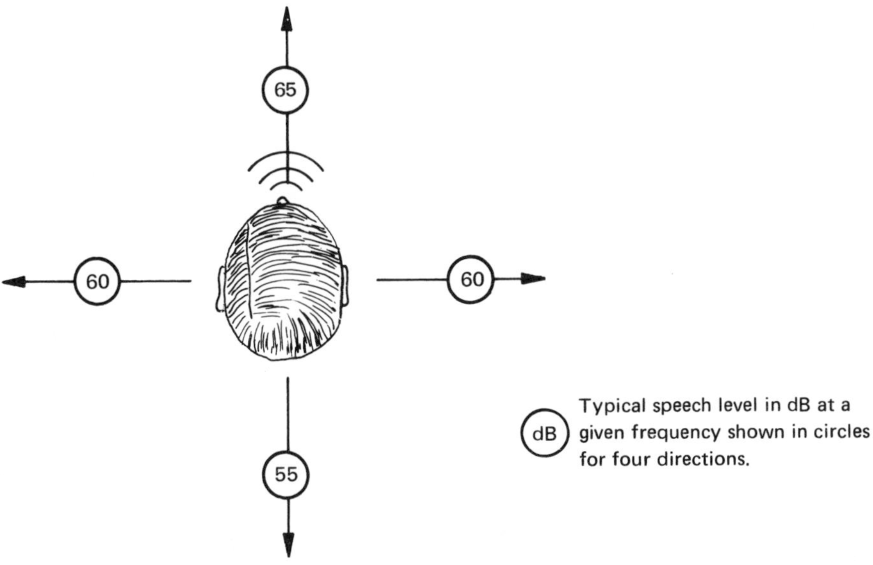

SPEAKER

Speaker orientation can be an important factor in open plans. As indicated on the above sketch, there is about a 10 dB difference in speech level between the front and rear of a speaker. Consequently, poor orientation of chairs or desks could contribute to unsatisfactory open plan privacy conditions.

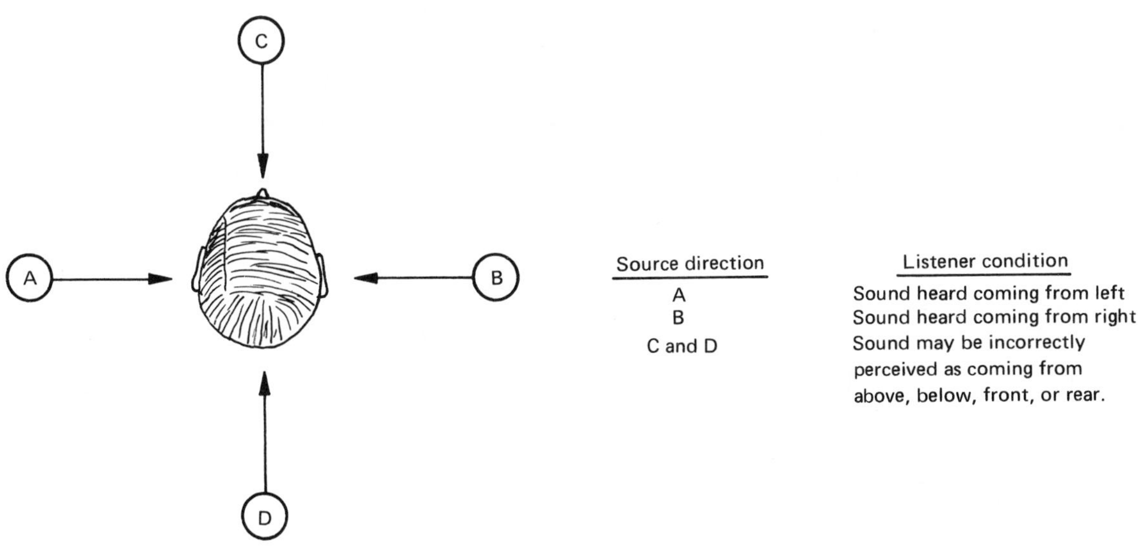

Source direction	Listener condition
A	Sound heard coming from left
B	Sound heard coming from right
C and D	Sound may be incorrectly perceived as coming from above, below, front, or rear.

LISTENER

Listener orientation affects perception of source direction, but not of the decibel level of the source.

SPEECH PRIVACY: Case Histories

Thirty-seven actual case histories of speech privacy are depicted below. The number in the circles represents the total case histories at an average TL. The study includes over 400 pairs of rooms in offices, hospitals, dormitories, motels, etc. It can be seen that no clear trend relating TL to occupant satisfaction exists. This type of scheme doesn't work, because there are other important acoustical factors that contribute to occupant satisfaction, e.g., background noise levels, how loud people talk, degree of privacy desired, etc. (See page 109 for a summary of the speech privacy factors.)

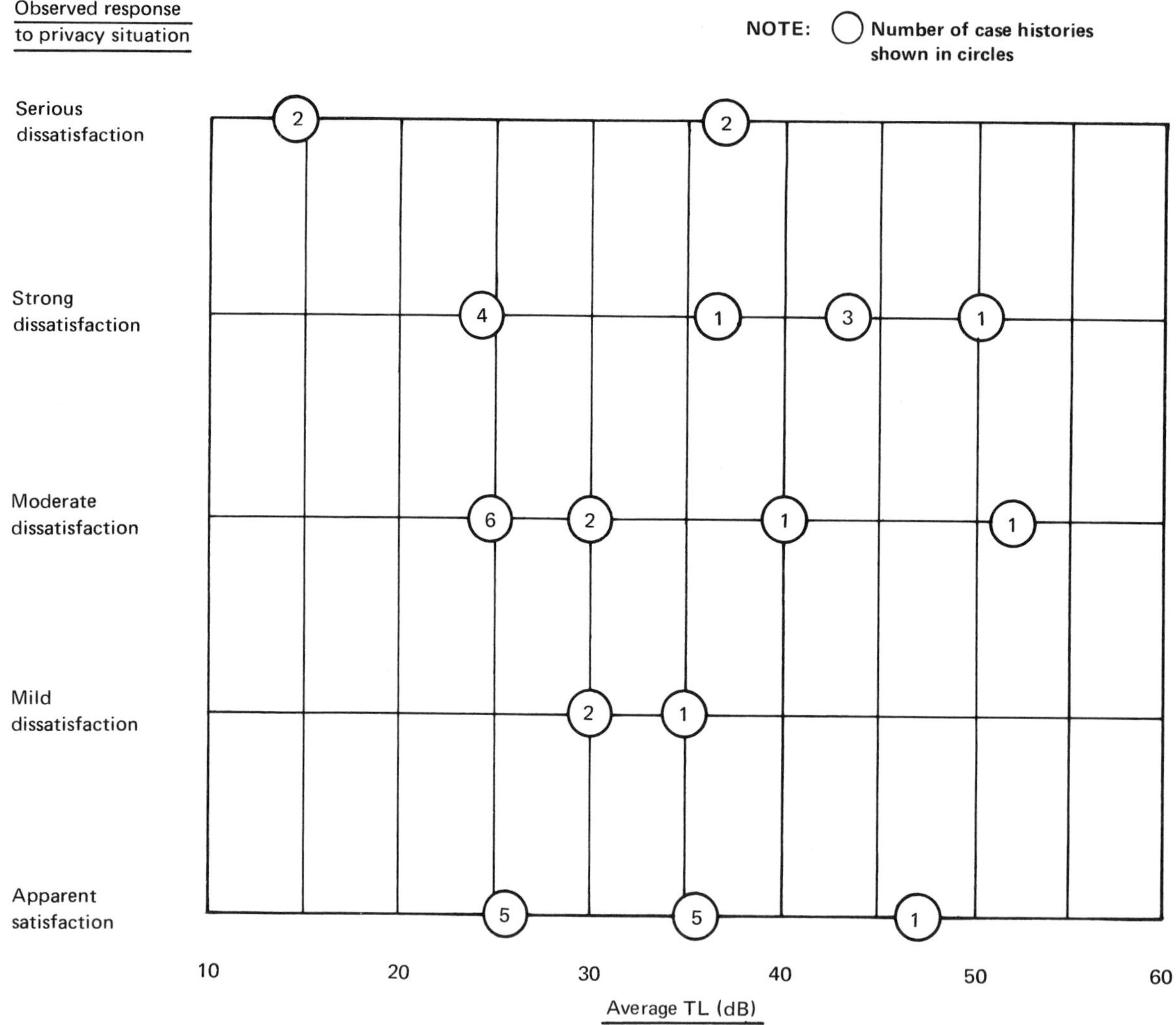

Ref.: W. J. Cavanaugh, W. R. Farrell, P. W. Hirtle and B. G. Watters, Speech Privacy in Bldgs., J. Acoust. Soc. Amer., Vol. 34, No. 4, April, 1962.

SPEECH PRIVACY: Background Noise

In speech privacy situations, annoyance is usually determined by the amount intruding speech is above (or below) the acceptable background noise rather than by the actual level of the intruding speech. This can be seen by the people annoyance curve below.

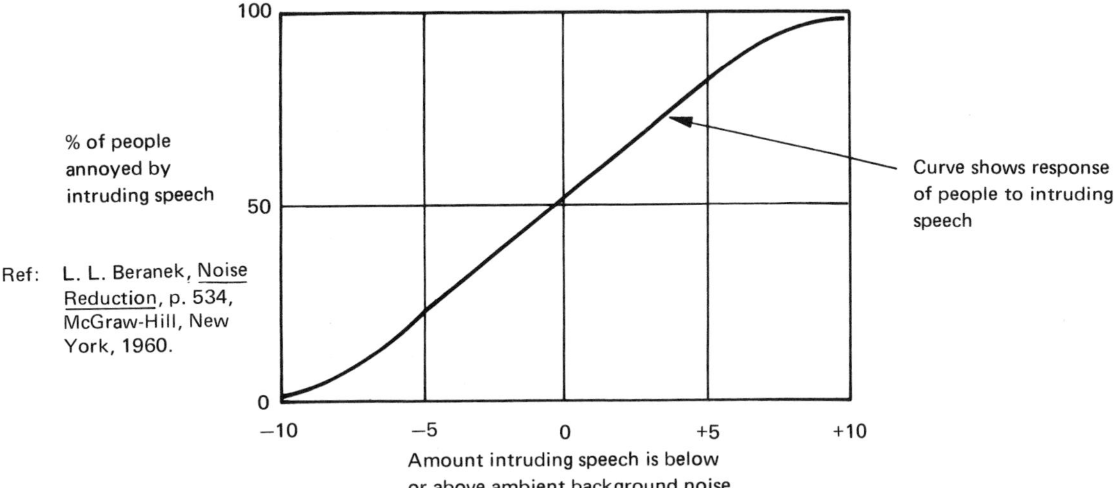

Controlled quiet from the mechanical system's air-distribution terminal devices can provide minimum background required to mask objectionable intruding noises in typical offices, as shown by the example below. Note that offices having "silent" systems will have serious complaints due to the lack of sufficient background noise levels.

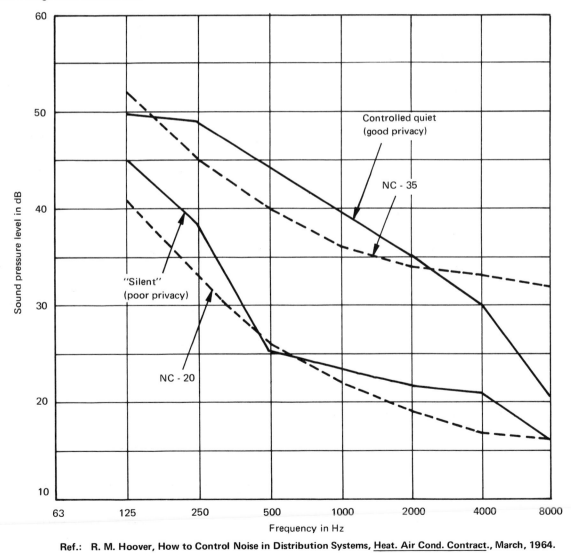

Sound pressure level in dB

60
50
40
30
20
10

Controlled quiet (good privacy)

NC - 35

"Silent" (poor privacy)

NC - 20

63 125 250 500 1000 2000 4000 8000

Frequency in Hz

Ref.: R. M. Hoover, How to Control Noise in Distribution Systems, Heat. Air Cond. Contract., March, 1964.

SPEECH PRIVACY: Analysis Sheet

Anticipated response to privacy situation

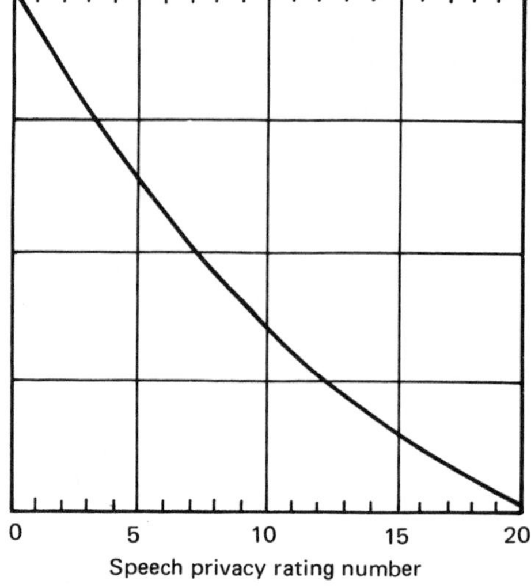

NOTE: Curve shows average response of people to intruding speech based on rating number figured below.

Speech rating

	Scale	A.	B.	C.
1. Speech effort – how people talk in source room	Loud 72 — Raised 66 — Conversational 60	____	____	____
2. Source room floor area (A_1) – approximates effect of source room absorption	125 → 10, 250 → 6, 500 → 3, 1000 (sq ft) → 0	____	____	____
3. Privacy allowance – degree of privacy desired	Confidential 15 — Normal 9	____	____	____
● Speech rating total		____	____	____

Isolation rating

	Scale	A.	B.	C.
4. Sound transmission class (STC) – accounts for transmission loss of common barrier		____	____	____
5. Noise reduction factor (A_2/S) – approximates effect of receiving room sound absorption and common barrier size	1 → -2, 0, 2, 3, 5 → 4, 5, 6, 7, 10 (unitless) → 8	____	____	____
6. Adjacent room background noise level (dBA) – masking sound available		____	____	____
● Isolation rating total		____	____	____

(Examples: A., B., C.)

Speech privacy rating number

Find speech privacy rating number by subtracting isolation rating total from speech rating total. Then use graph at top of sheet to predict degree of satisfaction.

A. ______________________________

B. ______________________________

C. ______________________________

GUIDELINES FOR USING THE SPEECH PRIVACY ANALYSIS SHEET

In many buildings the most important acoustical problem is the intrusion of intelligible speech. Experience shows that if the level of intruding speech is reduced by the common barrier to just the point where it cannot be understood, i.e., to a low speech articulation index between 0.05 to about 0.15, most occupants will be satisfied. The speech privacy analysis method can be used to design offices, hotels and motels, hospitals, dormitories, and apartment buildings and can handle most of the day-to-day privacy situations. However, very large rooms and spaces with electronically amplified speech or music are examples which fall outside the scope of the method. The step-by-step procedure outlined below is keyed to the rating factors on the speech privacy analysis sheet.

1. **Speech Effort**: Describes how people will talk in the source room.

   Conversational: The situation in most private offices, hotel rooms, hospital rooms, etc.

   Raised: In board rooms and conference rooms people usually increase their speech effort to a raised voice level.

   Loud: Exceptional cases such as psychiatrists' offices, where patients may sometimes become excited, or noisy business-machine rooms, where operators must speak in a loud voice to communicate.

   It is assumed that both the talker and the listener are located at least 2 to 3 ft away from the common barrier.

2. **Source Room Floor Area A_1**: The source room size A_1 (floor area in square feet) is important because it approximates the effect of room sound absorption.

   In a small room the sound will reflect more frequently from the room surfaces and will result in a buildup of sound intensity. Conversely, in a large room the sound will tend to spread out and the intensity level of the speech signals will be less. It is assumed by the speech privacy method that sound-absorbing ceiling treatment is provided as standard practice for today's office buildings and other spaces where speech privacy is important.

3. **Privacy Allowance**: Deals with the kind of privacy which is desired.

   Normal: Means the occupant wants freedom from disturbing intruding speech. For example, even engineers or salesmen who work closely together would want protection from their neighbors' distracting conversations.

   Confidential: The occupant doesn't want his conversation overheard in next room. Doctors and lawyers usually require confidential privacy. Executives and supervisors often need this degree of privacy to be free to discuss employee relations, costs and fees, etc.

4. **Sound Transmission Class STC**: Accounts for sound transmission loss of common barrier.

   The STC is a single-number rating of a barrier's airborne sound transmission performance over a standard frequency range. STC ratings are shown on pages 73 through 75 for various building constructions. Note that if all other speech privacy factors are known, one can solve directly for the required STC by setting the speech privacy rating number equal to 0, which will not allow excess intruding speech.

5. **Noise Reduction Factor A_2/S:** Approximates the effect of receiving room sound absorption and common barrier size.

   The receiving room size A_2 (floor area in square feet) is important because noise buildup is greater in small, reverberant rooms than in well-treated rooms. The common barrier size S (surface area in square feet) is also an important factor since it will be the sound source in the receiving room.

6. **Adjacent Room Background Noise Level (dBA):** Masking sound available.

   The adjacent room background noise levels should be designed to mask or "cover up" the intruding speech signals. Background noise should be bland and continuous so that people will hardly notice it. Recommended background NC levels are presented on page 86. (Remember dBA values are about 10 dB greater than corresponding NC levels.) It is also important that the source of background noise be reliable. For example, in offices where the work activity fluctuates, the noise produced by the activity will also fluctuate. The designer should always specify reliable background sources such as air-flow noise at mechanical system registers, or in special situations neutral noise (*not* music) from electronic noise generators.

References

1. W. J. Cavanaugh; W. R. Farrell; P. W. Hirtle; and B. G. Watters: Speech Privacy in Buildings, *J. Acoust. Soc. Amer.*, Vol. 34, no. 4, April, 1962.
2. "Speech Privacy Design Analyzer," Owens-Corning Fiberglas Corp., Toledo, Ohio, 1962.
3. R. W. Young: Re-vision of the Speech Privacy Calculation, *J. Acoust. Soc. Amer.*, Vol. 38, no. 4, October, 1965.

SPEECH PRIVACY: Design Examples

Supervisor's office, adjacent to conference room, for insurance company. Carpeted floors and acoustical tile ceilings. People in the conference room can be expected to often talk in a "raised" voice to span the distance of the conference table. Both rooms will be used at times for work of a "confidential" nature.

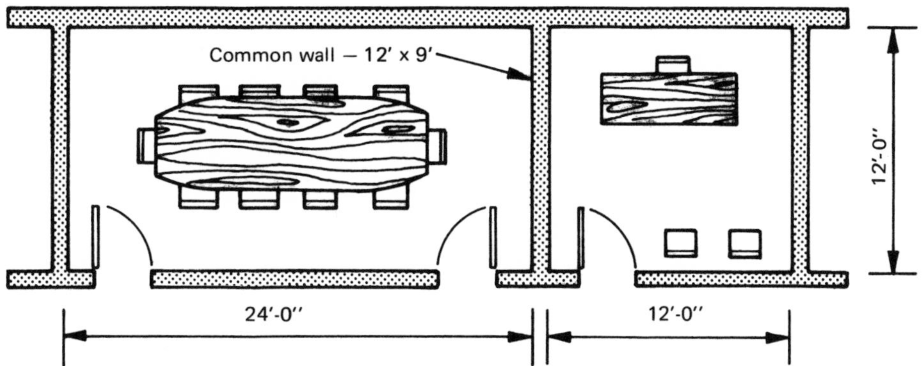

Source room floor area:	A_1 = 24 X 12 = 288 sq ft (conf. room)
Adjacent room floor area:	A_2 = 12 X 12 = 144 sq ft (office)
Noise reduction factor:	A_2/S = 144/108 = 1.3
Minimum background levels	NC-30 or about 40 dBA in supervisor's office
(See page on recommended NC's):	NC-25 or about 35 dBA in conference room

EXAMPLES (Use analysis sheet on following page to find required common wall)

A. Conference room to supervisor's office:
Try STC 38 const.: 2 x 4 wood studs 16" o.c. w/5/8" gyp. bd. on both sides and 1/4" hardwood panels on one side.

Strong dissatisfaction

B. Conference room to supervisor's office:
Use STC 49 const.: 2 x 4 staggered wood studs 24" o.c. w/5/8" gyp. bd. on both sides and 1/4" hardboard on both sides.

Apparent satisfaction

C. Supervisor's office to conference room:
Note that source room area (A_1) now is 144 sq ft and A_2/S will be 288/108 = 2.6. Also, "conversational" speech is assumed since the office is reasonably small. Although STC 48 is required, use STC 49 construction from Example B since most stringent requirement should dictate.

Apparent satisfaction

SPEECH PRIVACY: Analysis Sheet

Anticipated response to privacy situation

NOTE: Curve shows average response of people to intruding speech based on rating number figured below.

Speech rating

1. Speech effort — how people talk in source room

2. Source room floor area (A_1) — approximates effect of source room absorption

3. Privacy allowance — degree of privacy desired

Isolation rating

4. Sound transmission class (STC) — accounts for transmission loss of common barrier

5. Noise reduction factor (A_2/S) — approximates effect of receiving room absorption and common barrier size

6. Adjacent room background noise level (dBA) — masking sound available

	Examples A.	B.	C.
1. Speech effort	66	66	60
2. Source room floor area	6	6	9
3. Privacy allowance	15	15	15
● Speech rating total	87	87	84
4. Sound transmission class	38	49	48
5. Noise reduction factor	-2	-2	1
6. Adjacent room background noise level	40	40	35
● Isolation rating total	76	87	84
Speech privacy rating number	11	0	0

Speech privacy rating number

Find speech privacy rating number by subtracting isolation rating total from speech rating total. Then use graph at top of sheet to predict degree of satisfaction.

A. Conference room to supervisor's office

B. Conference room to supervisor's office

C. Supervisor's office to conference room

STC RATINGS FOR ENCLOSED OFFICES

The following table shows minimum STC ratings required between typical enclosed office spaces. It assumes "average" or conversational levels of speech in the source room, and office dimensions of 12 by 12 ft by 9 ft ceiling height. STCs are given for both normal and confidential occupant privacy.

STC RATINGS

Office Types	Background Noise Level, dBA	Normal Privacy	Confidential Privacy
Very quiet offices	30	50	56
Quiet offices	40	40	46
Moderately noisy offices	50	30	36

SPEECH PRIVACY: Open Plan Analysis Sheet

Open plan dimensions

Barrier

H(ft)

Acoustical "line of sight"

Source or listener

D(ft)

Source or listener

Examples

	A.	B.	C.

Speech rating

1. Speech effort – how people talk in room.

Loud 72 — Raised 66 — Conversational 60 — Low 54

2. Privacy allowance – degree of privacy desired.

Confidential 15 — Normal 9

Isolation rating

● Speech rating total

3. Distance from source to listener – Table approximates effect of room sound absorption and sound level falloff with distance (D) from source to listener.

Room finishes		Distance D, ft.					
Ceiling	Floor	3	6	12	25	50	100
Reflecting	Reflecting	0	3	6	9	12	15
Reflecting	Absorbing	0	4	8	12	16	20
Absorbing	Reflecting	0	5	10	15	20	25
Absorbing	Absorbing	0	6	12	18	24	30

4. Partial – height barrier – Table accounts for attenuation from barrier with ceiling absorption based on NRC of 0.75. Barrier width should be at least twice its total height.

Barrier height H – Portion above acoustical "line-of-sight" in feet.	Distance D, ft					
	3	6	12	25	50	100
1	11	7	4	2	0	0
2	14	10	7	4	3	2
3	15	11	8	5	4	3
4	16	12	9	6	5	4

5. Room background noise level (dBA) – Masking sound available.

Speech privacy rating number

● Isolation rating total

Find speech privacy rating number by subtracting isolation rating total from speech rating total.
Satisfactory conditions are anticipated when speech privacy rating number is 0 or less.
(Use graph on p. 108 to assess approx. degrees of satisfaction.)

A. ______

B. ______

C. ______

GUIDELINES FOR USING THE SPEECH PRIVACY ANALYSIS SHEET (OPEN PLAN)

In the open plan space one of the most important design problems is to provide acoustical privacy from the occupants' intelligible speech. The speech privacy analysis method can be used to evaluate and design offices, schools, banks, etc. and can handle most of the day-to-day open plan privacy situations. The method assumes that the length and width of the open plan space are large compared to its height. However, open plans with subareas largely enclosed by partial-height barriers will have the acoustical characteristics of small rooms and therefore fall outside the scope of the method. The step-by-step procedure outlined below is keyed to the rating factors on the open plan speech privacy analysis sheet.

1. **Speech Effort**: Describes how people will talk in the room.

   Low: The situation in most open plan spaces that are heavily treated with sound absorbing materials. (A whispering voice, for example, would fall below the low voice level.)

   Conversational: Persons in the open plan must often learn to speak in at least a conversational voice level so they will not disturb their neighbors.

   Raised: In conference situations people can be expected to talk in a raised voice level to span the distance of conference tables.

   Loud: Workers in noisy business-machine spaces must speak in a loud voice to communicate.

2. **Privacy Allowance**: Deals with the kind of privacy which is desired.

   Normal: Means the occupant wants freedom from disturbing intruding speech. Generally, select normal privacy in design for most open plan privacy situations.

   Confidential: The occupant doesn't want his conversation overheard. Banks, financial offices, etc., are examples that usually require confidential privacy. Executives, supervisors, and counselors often need this degree of privacy to be able to converse with a select individual or group without being overheard.

3. **Distance from Source to Listener D**: Accounts for the attenuation of voice voice levels with distance.

   This factor accounts for the attenuation in the voice level of the source at the position of the nearest listener, a distance D away in feet. Use the smallest value of D measured between people who want privacy between each other. Assume also that people who wish to converse will be fairly close together. Note that conditions may range from a room with a hard, sound-reflecting ceiling and floor to one with a sound-absorbing ceiling and a deep carpeted floor. Consequently, attenuation with distance ranges from 3 to 6 dB per doubling of distance as shown by the values in the table.

4. **Partial-height Barrier:** Accounts for additional attenuation of partial-height barrier.

The ceiling is assumed to be covered with a sound absorbing material having a NRC of at least 0.75. Also, it is assumed that the barrier is located midway between the talker (source) and listener – the least effective location. The barrier height H is the portion of the barrier above the acoustical line of sight between the talker and listener in feet. Note that the barrier width should be at least twice its total height to prevent flanking of speech signals around its ends.

5. **Room Background Noise Level (dBA):** Masking sound available.

Room background noise levels should be designed to mask or "cover up" the intruding speech signals. Background noise should be bland and continuous so people will hardly notice it, and be uniform over the entire open plan space as well. Recommended ranges for background NC-levels are presented on page 86. (Remember dBA values are about 10 dB greater than the corresponding NC-level.) One method of producing reliable background noise in the open plan space is to provide loudspeakers distributed above the sound absorbing panels of a suspended ceiling.

References

1. L. L. Beranek, *Noise and Vibration Control*, McGraw-Hill, New York, 1971.
2. P. W. Hirtle; B. G. Watters, and W. J. Cavanaugh; Acoustics of Open Plan Spaces – Some Case Histories, *J. Acoust. Soc. Amer.*, Vol. 46, no. 91A, 1969.
3. R. Pirn: Acoustical Variables in Open Planning, *J. Acoust. Soc. Amer.*, Vol. 49, no. 5, May, 1971.

SPEECH PRIVACY: Background Noise for Open Plan Spaces

The background noise in the open plan should have a neutral tonal quality. Shown below is a range for background noise suggested by Beranek et al. that will provide a pleasant acoustical background. Note that the dBA or NC-level of the background will depend on the specific masking requirements.

Ref.: L. L. Beranek, W. E. Blazier and J. J. Figwer, "Preferred Noise Criterion (PNC) Curves and Their Application to Rooms," *J. Acoust. Soc. Amer.*, Vol. 50, no. 5, November , 1971.

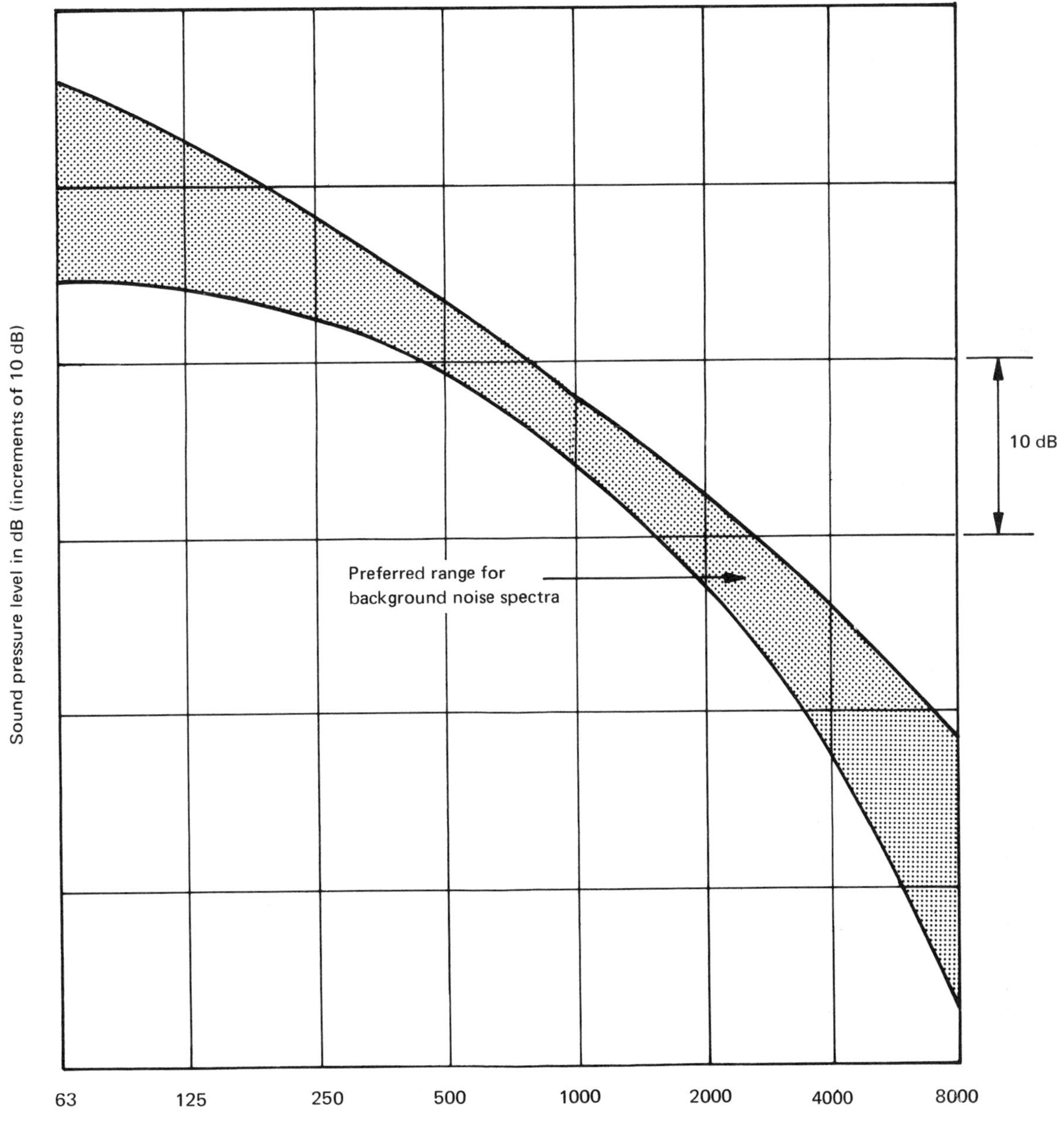

NOTE: **Slightly higher SPL's at both low and high frequencies may be permitted in some air-conditioned buildings.**

SPEECH PRIVACY: Communication in the Open Plan

Background noise levels in the open plan space must be high enough to provide satisfactory speech privacy conditions, and may therefore interfere with speech communication. The graph below can be used to estimate the maximum distance people may be separated and still converse satisfactorily at a given voice level. For example, in an open plan having a background level of 55 dBA, a high level of about NC-45, two people can converse with "normal" speech effort up to a distance of 9 1/2 ft (see dashed lines). If conversations or lectures to large groups cannot be restricted to the distance from the graph, they must be held in enclosed spaces to avoid loss of speech privacy for other room occupants.

Ref.: L. L. Beranek, *Acoustics*, p. 420, McGraw-Hill, New York, 1954.

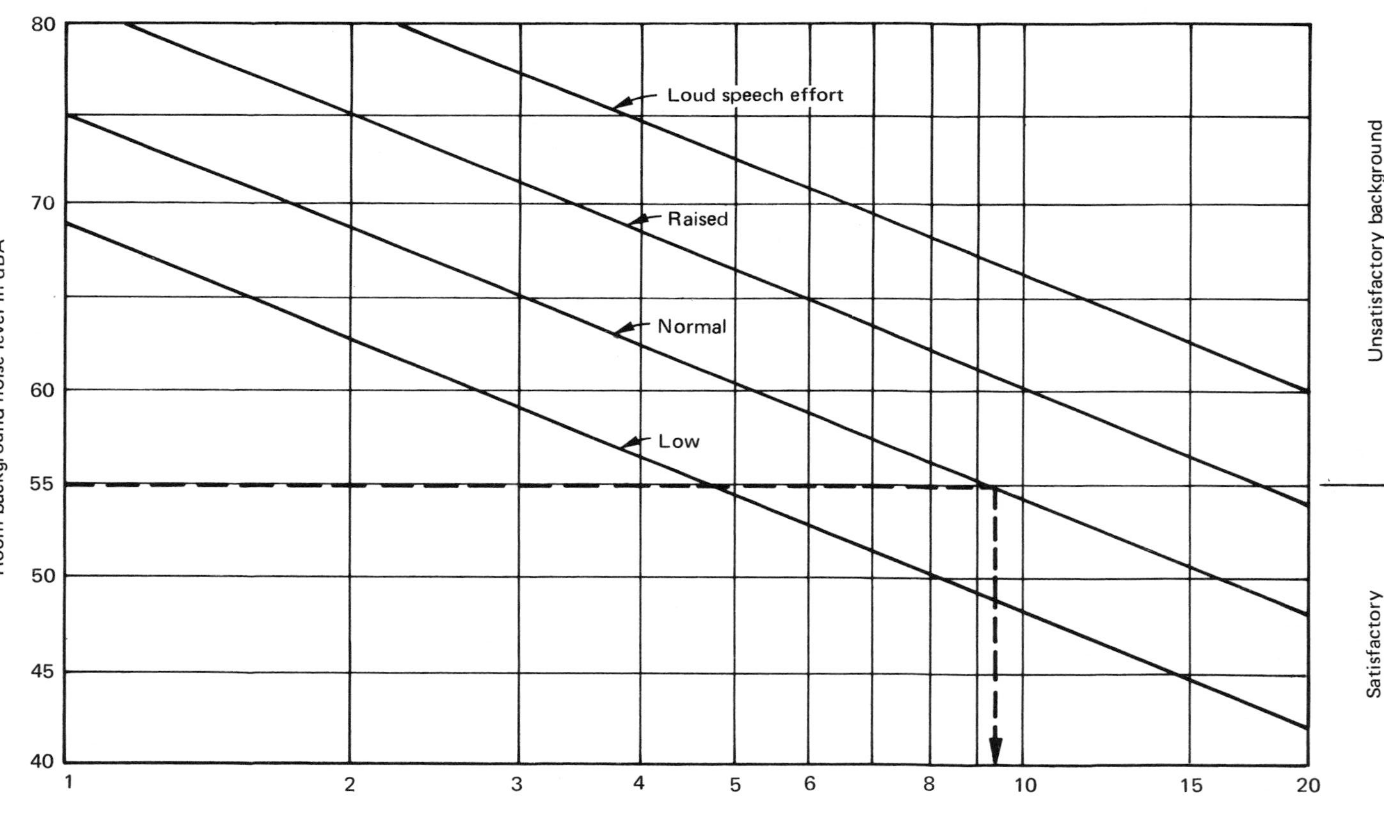

NOTE: The graph is based on tests of average male voices with talker and listener facing each other, using unexpected word material.

SECTION 5

Mechanical System Noise and Vibrations

MECHANICAL SYSTEM NOISE AND VIBRATIONS: Vibrating Equipment

The vibration produced in buildings by mechanical equipment can be felt and heard by building occupants. Vibrations can travel through solid members such as columns and beams, which may reradiate sound at great distances. It is therefore important that the vibrating equipment be properly isolated from the building structure with resilient mounts. If possible, vibrating equipment should be positioned away from the center of floor spans, near columns or load-bearing walls where they will have better structural support.

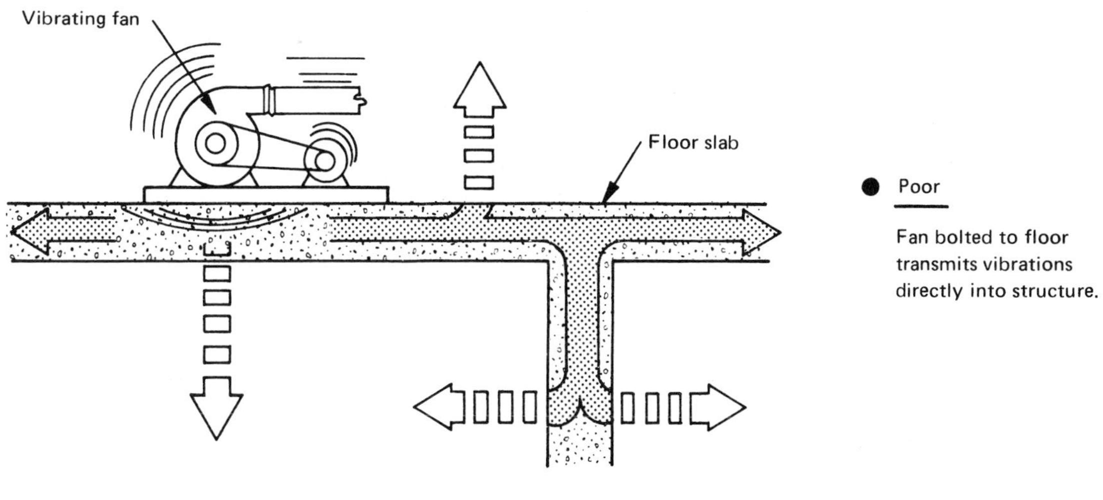

- Poor

Fan bolted to floor transmits vibrations directly into structure.

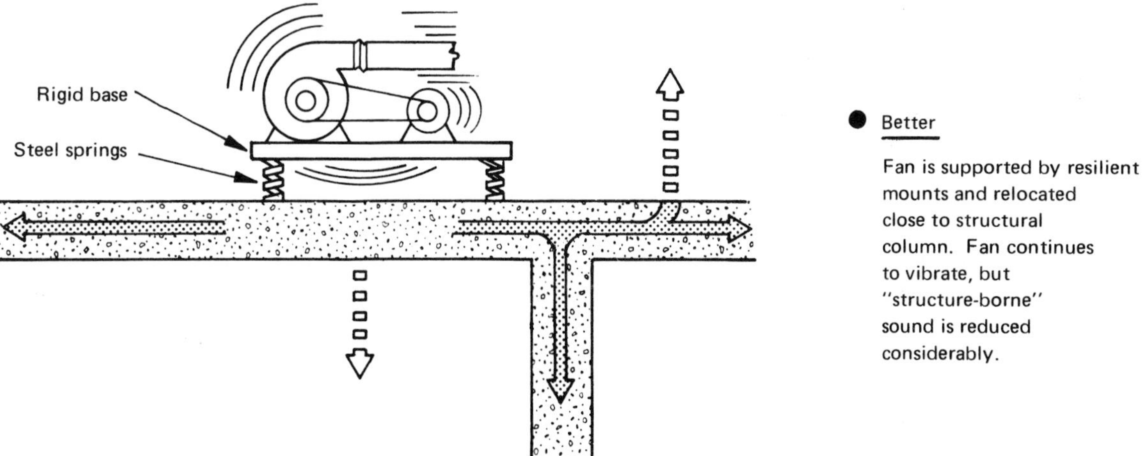

- Better

Fan is supported by resilient mounts and relocated close to structural column. Fan continues to vibrate, but "structure-borne" sound is reduced considerably.

MECHANICAL SYSTEM NOISE AND VIBRATIONS: Basic Vibration Theory

The principle of vibration isolation involves supporting the vibrating equipment by resilient materials such as (1) ribbed neoprene pads, (2) precompressed glass-fiber pads, and (3) steel springs. The goal is to choose the proper resilient material which will provide a natural frequency when loaded that is three or more times lower than the driving frequency of the equipment. Driving frequency is an operational characteristic of the equipment which should be obtained from the manufacturer. Natural frequency (f_n) in cycles per second (or Hertz) can be calculated since it varies with deflection as follows:

$$f_n = 3.13\sqrt{\frac{1}{Y}}$$

where f_n = natural frequency, Hz (cps)
Y = static deflection, in.

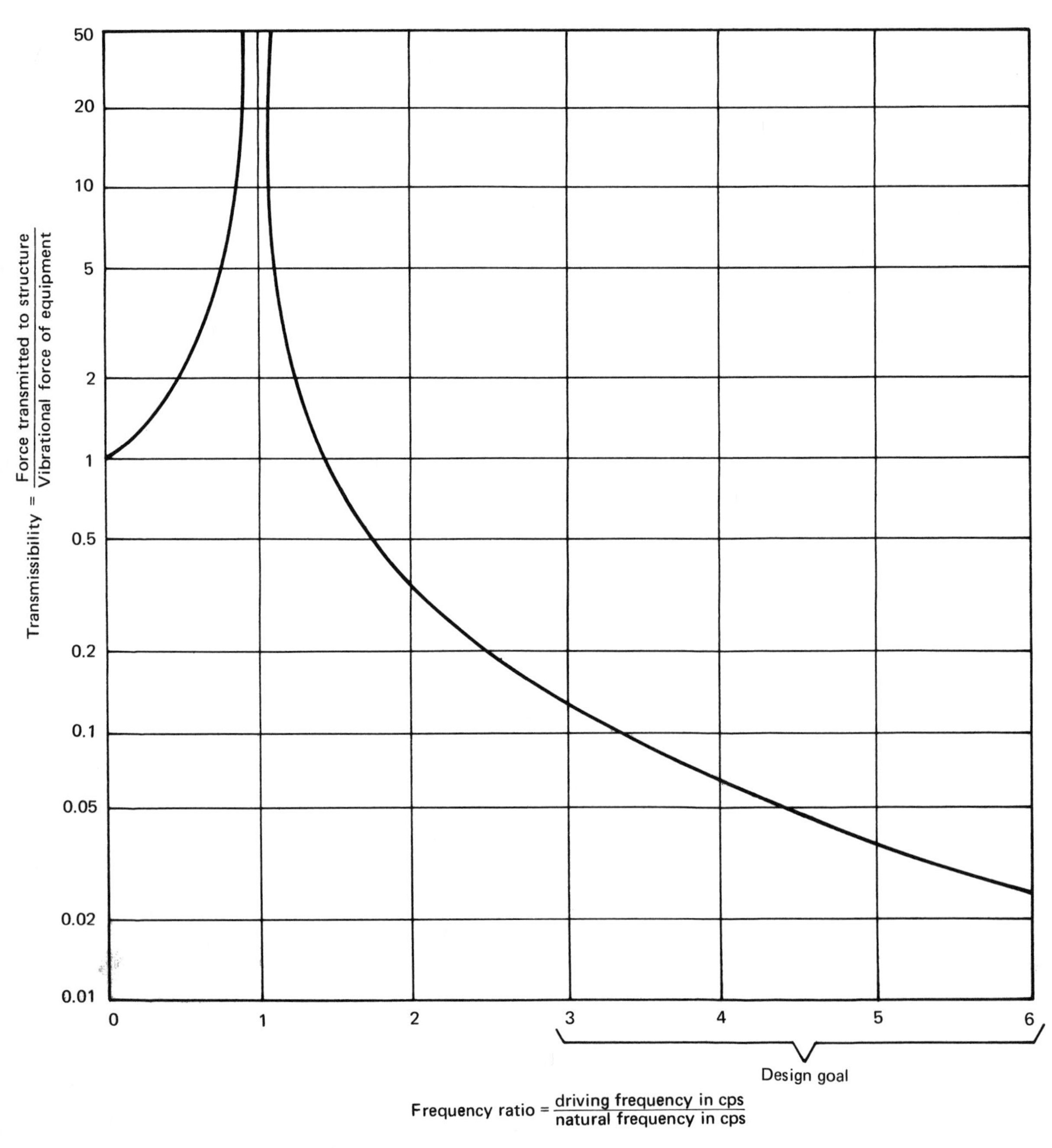

MECHANICAL SYSTEM NOISE AND VIBRATIONS: Natural Frequency and Deflection

Curves below show natural frequency (f_n) for (1) free-standing steel springs, (2) precompressed glass-fiber, (3) rubber, and (4) cork. The most resilient isolators are springs, because they have the largest deflections.

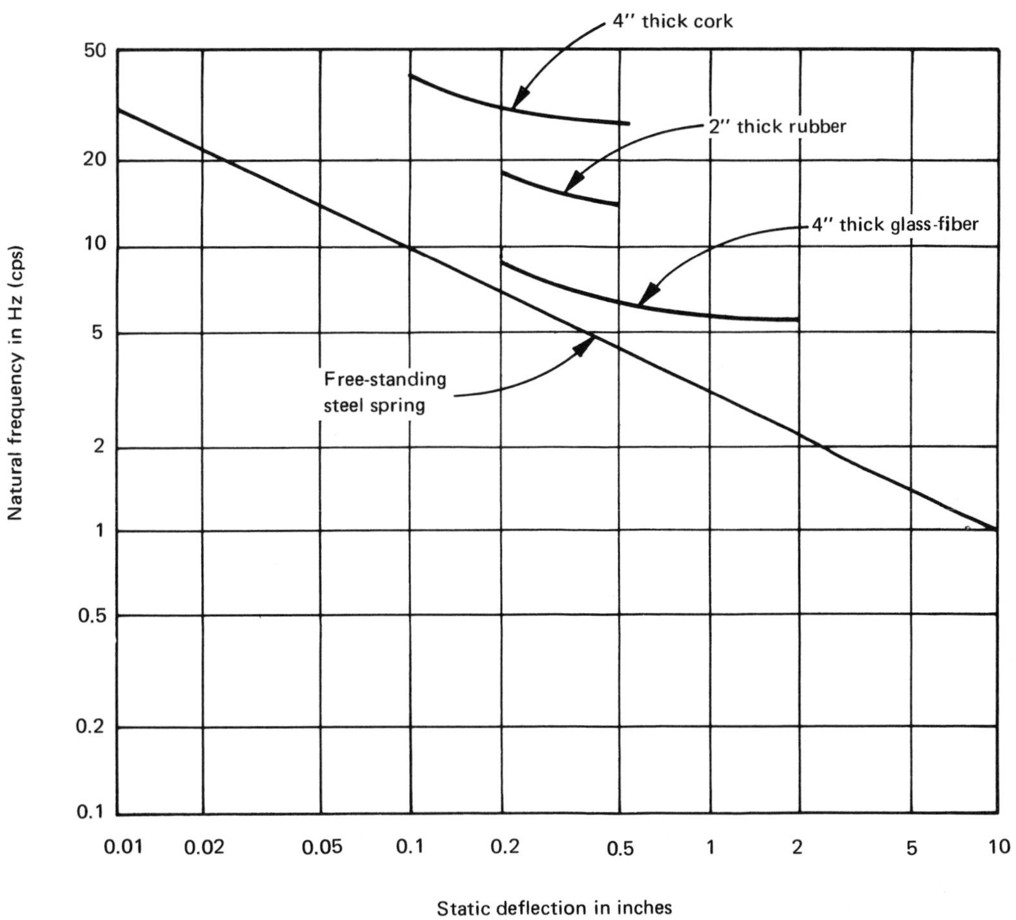

MECHANICAL SYSTEM NOISE AND VIBRATIONS: Vibration Isolation Design Graph

The curves below can be used to find the static deflection required for vibrating equipment isolators. For example, an air exhaust fan for a gymnasium that operates at a driving frequency of 520 cpm will require resilient isolators having a static deflection of at least 1 inch (see dashed lines on graph). The graph is based on the assumption that the resilient materials will be in turn supported by a rigid base as flexible supports (e.g., wood floor systems) require special consideration.

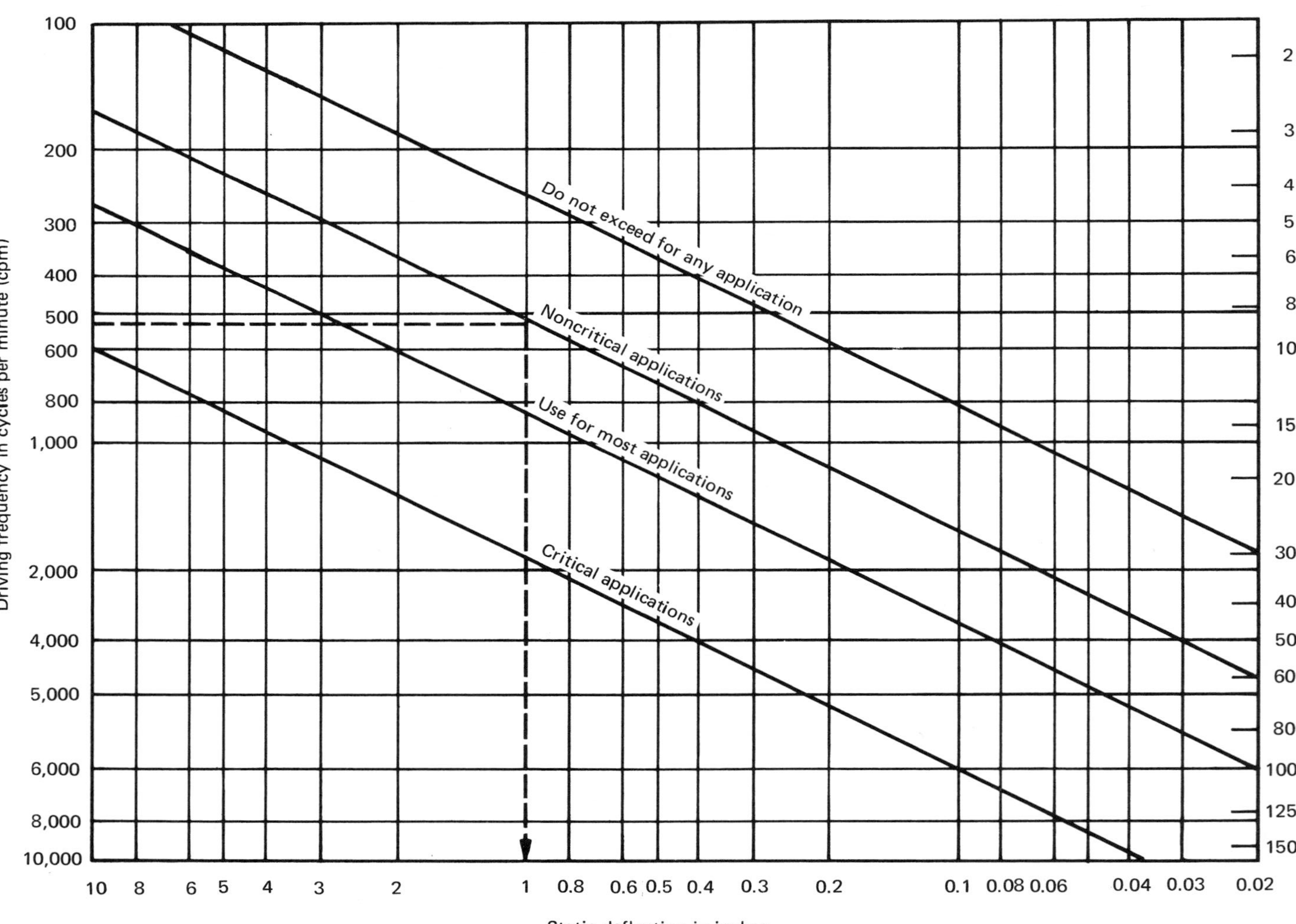

MECHANICAL SYSTEM NOISE AND VIBRATIONS: Mechanical Duct Systems

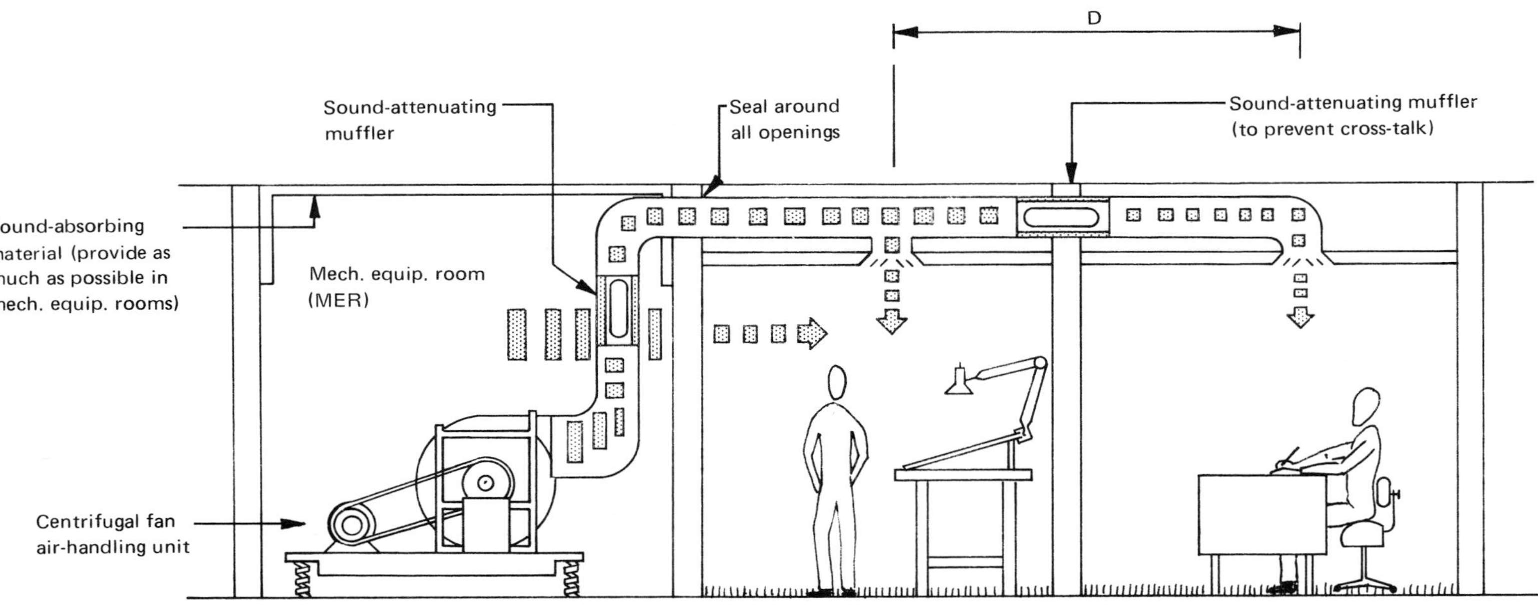

Duct-borne noise isolation

- Provide smooth duct turns
- Sound is attenuated in ducts by:
  1. Distance
  2. Bends and branches
  3. End reflection at openings into a room
  4. Internal glass-fiber lining and mufflers
- Avoid high air velocities in ducts

Cross-talk noise isolation

- Greater the distance D between room registers the better
- Line common ducts with internal glass-fiber or provide sound-attenuating mufflers

NOTE: **Mechanical duct isolation must be carefully checked in both the supply and the return air systems since sound travel in ducts is independent of air-flow direction! Refer to ASHRAE Guide and Data Book chapter, "Sound and Vibration," for duct-borne and cross-talk noise isolation computational procedures.**

MECHANICAL SYSTEM NOISE AND VIBRATIONS: Mechanical Equipment Room Treatment

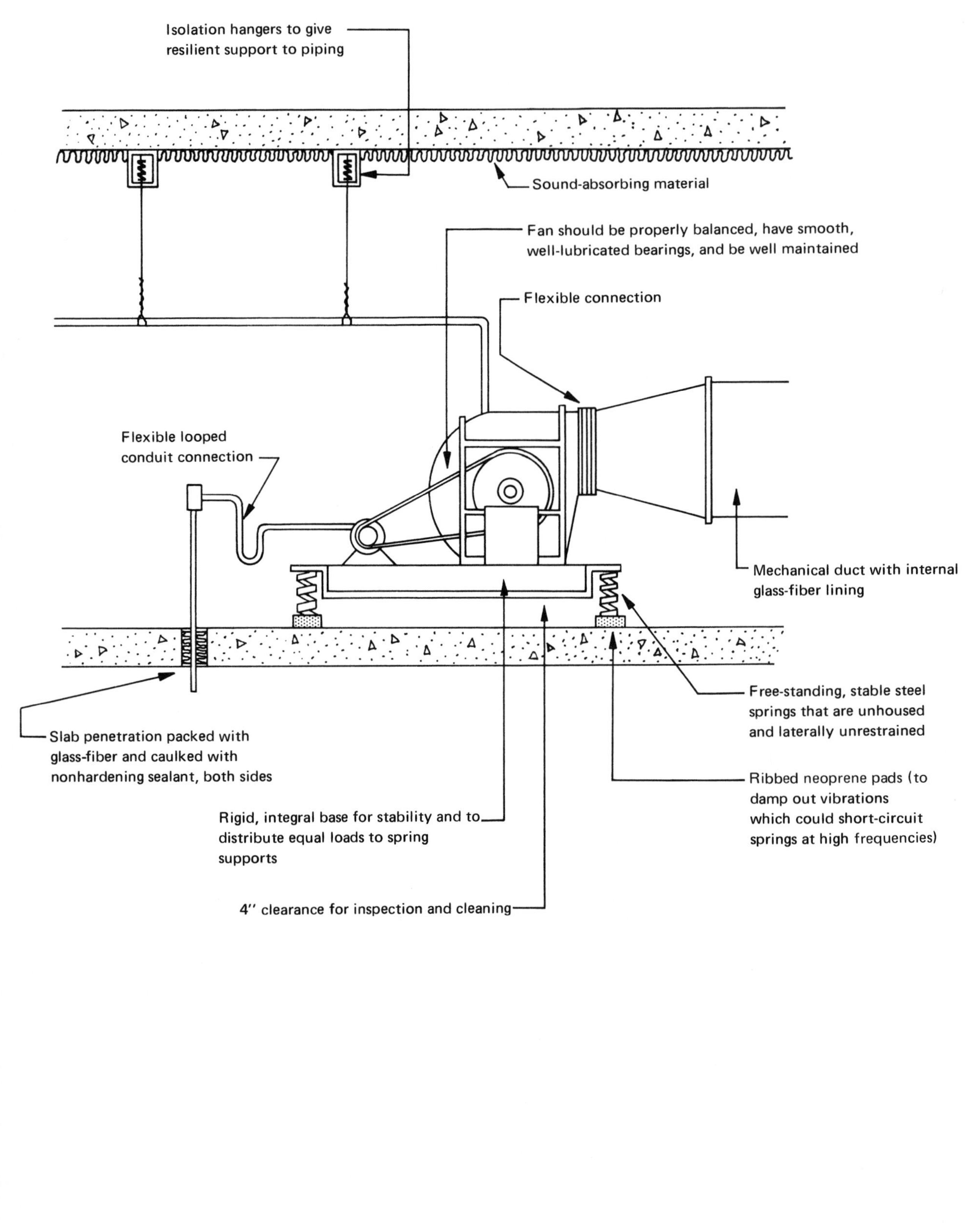

VIBRATION ISOLATION GUIDELINES

1. **Fans:** Large centrifugal fans can be a source of considerable low frequency vibration. These fans and their motors should be mounted on a common, rigid base to avoid misalignment, which wears out fan belts and bearings. The rigid base is in turn placed on spring isolators. Sometimes the base is a concrete slab called an "inertia block." This inertia block evens out the load on the springs besides providing a rigid base. Additionally, its mass reduces the amplitude of vibration, but not the forces, and contributes somewhat to the airborne sound transmission loss through the floor slab below.
2. **Refrigeration Compressors:** Large, low-speed reciprocating compressors should be isolated by springs and inertia blocks. High-speed centrifugal compressors require less isolation and often can be isolated properly with several layers of ribbed neoprene.
3. **Cooling Towers:** Cooling tower vibration involves the low frequency vibration of propeller-type fans as well as the high frequency waterfall noise. High-deflection steel springs and ribbed neoprene mounts are required.
4. **Transformers:** Transformers should be supported with several layers of ribbed neoprene. Also, partial or demountable enclosures, lined with sound-absorbing materials, can be used to help isolate annoying humming noises.

References

1. *ASHRAE Guide and Data Book: Systems* (1970) published periodically by the American Society of Heating, Refrigerating, and Air-Conditioning Engineers, Inc. (ASHRAE), 345 East 47th Street, New York, N. Y. 10017.
2. R. E. Fischer: Some Particular Problems of Noise Control, *Architect. Rec.*, September, 1968.

MECHANICAL SYSTEM NOISE AND VIBRATIONS: Cross-talk

Untreated mechanical ducts can act like speaking tubes to transmit unwanted sound from one room to another, as shown below. The sound transmission loss through common ducts should equal the common wall transmission loss. Use duct linings or sound-attenuating mufflers where required.

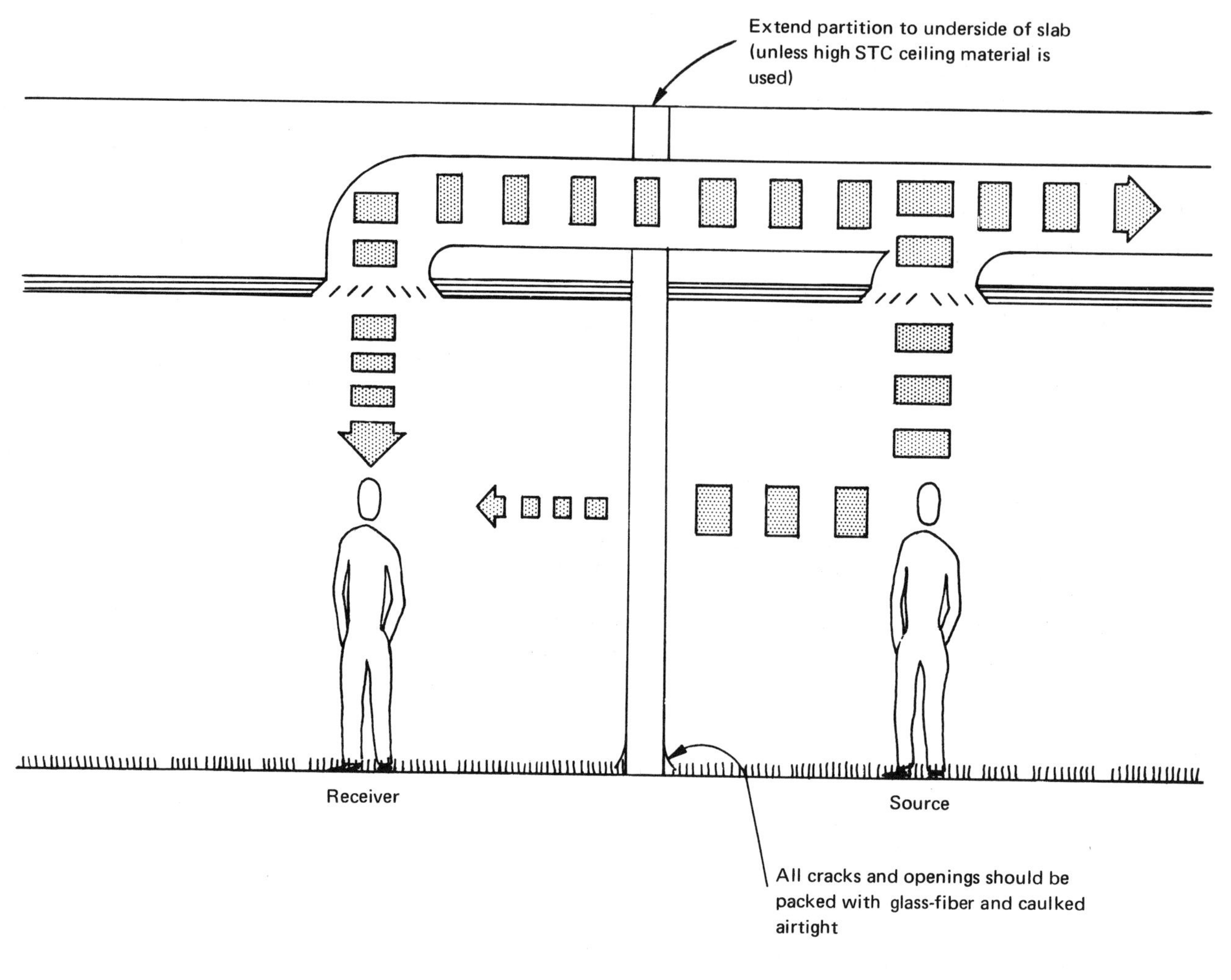
Extend partition to underside of slab (unless high STC ceiling material is used)
Receiver
Source
All cracks and openings should be packed with glass-fiber and caulked airtight

NOTE: Ductwork should not be rigidly connected to partitions since they may transmit mechanical vibrations.

MECHANICAL SYSTEM NOISE AND VIBRATIONS: Glass-fiber Duct Linings

Sound attenuation with glass-fiber linings is often required in mechanical ducts to prevent transmission of equipment noise as well as cross-talk between rooms. The curves below show attenuation in decibels per foot of duct length for internal glass-fiber linings. The curves show 1- and 2-in. thick linings for a 16-in. deep mechanical duct. Actual data should be used whenever possible. For example, the attenuation shown below would be less for larger ducts with identical linings.

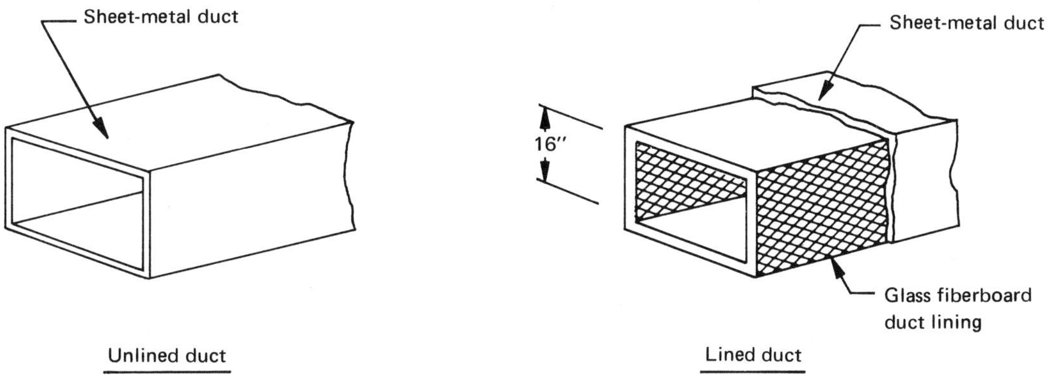

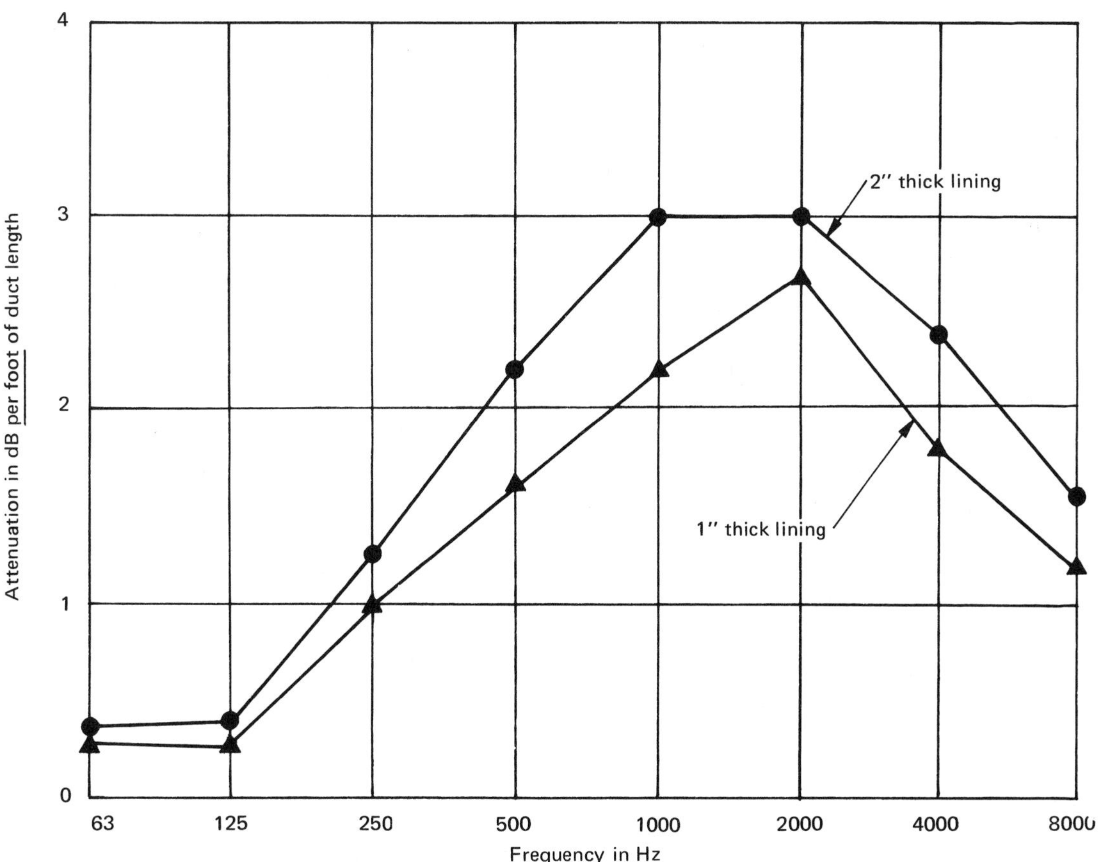

NOTE: Since ducts are usually made of thin sheet metal (or lightweight glass fiberboard), sound can easily pass through the duct wall. Often an outer cover of plaster may be needed to improve isolation where ducts run through noisy spaces.

MECHANICAL SYSTEM NOISE AND VIBRATIONS: Sound-attenuating Mufflers

Prefabricated sound-attenuating mufflers are especially useful where high attenuation over a wide frequency range is required and the available length of duct is limited. They are normally available in both rectangular and round form. Typical sound-attenuating data for a 3-ft long muffler are shown below. Refer to manufacturer's test data for anticipated performance at actual design conditions.

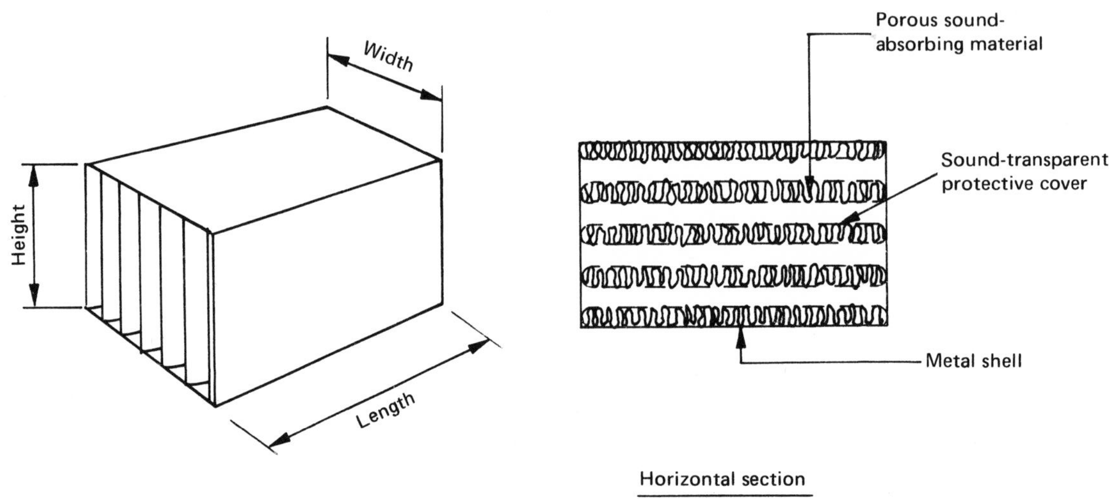

Rectangular muffler

Sound Attenuation in dB:

Frequency (Hz)	125	250	500	1000	2000	4000	8000
Attenuation (dB) 3-ft long muffler	13	28	36	39	40	39	27

NOTE: **Also, check manufacturer's test data for the noise generated by the muffler itself. Air-flow through mufflers produces sound (from turbulence) which may be critical in applications where low acoustical backgrounds are required.**

RECOMMENDED AIR SPEEDS AT ROOM REGISTERS

The following table shows approximate air velocities in feet per minute (fpm) required to generate background noise levels at room registers. It assumes internal glass-fiber duct linings and at least 1/2-in. wide register slot openings. The table values should be used for design purposes only when specific data on manufacturers' equipment are unavailable.

Range of Noise Criteria	Supply Register (fpm)	Return Grille (fpm)
NC-20 to NC-25	300 to 350	360 to 420
NC-25 to NC-30	350 to 425	420 to 510
NC-30 to NC-35	425 to 500	510 to 600
NC-35 to NC-40	500 to 575	600 to 690
NC-40 to NC-45	575 to 650	690 to 780

MECHANICAL SYSTEM NOISE AND VIBRATIONS: Impact Noise

Typically, impact noises are erratic and can be caused by (1) hard heel footfall, (2) dropped objects, (3) shuffled furniture, etc. Impacts on floors, as shown on the sketches at the bottom of the page, are radiated directly downward. However, they also can be transmitted through the structure and radiated at distant locations. Impact noise test results using a standard tapping machine on three floor-ceiling constructions are shown below:

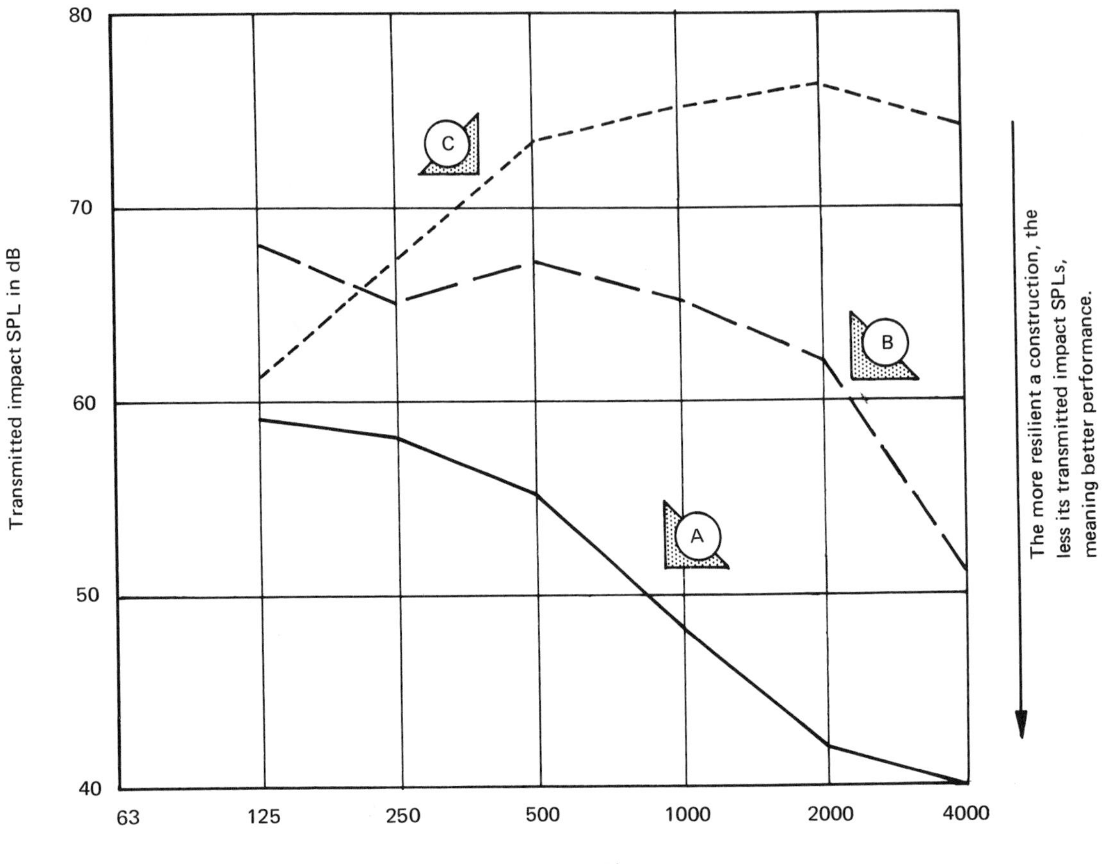

Frequency in Hz

	IIC Rating
A. 6″ reinf. conc. slab w/3/4″ T & G wood flooring nailed to 1 1/2″ x 2″ wood battens, 16″ o.c., floating on 1″ glass-fiber blanket (83 psf)	57
B. 4 3/8″ reinf. conc. slab w/7/8″ gyp. board suspended 4″ below w/wire hangers (62 psf)	47
C. 4″ reinf. conc. slab (54 psf)	25

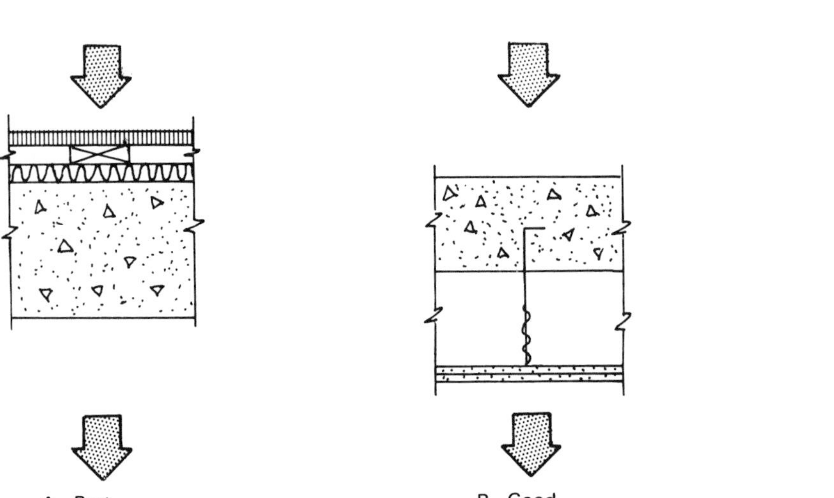
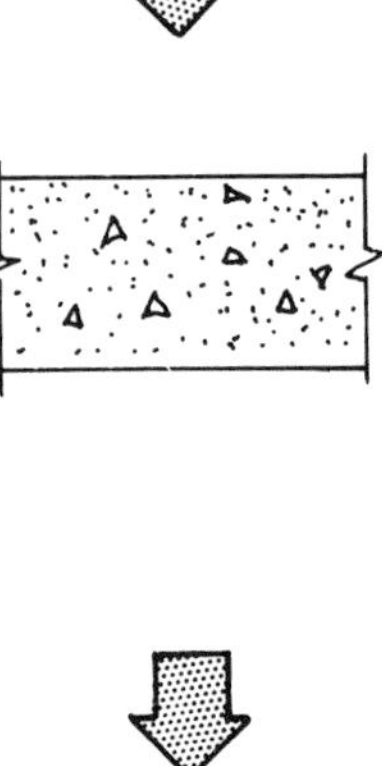

A. Best B. Good C. Poor

MECHANICAL SYSTEM NOISE AND VIBRATIONS: Floor-ceiling Constructions for Impact Isolation

Carpeting and resilient floor tiles such as rubber and cork can be used to cushion impacts. They are most effective, however, at middle and high frequencies (e.g., footfall "clicks"), while low frequency "thuds" still may pass through. To achieve high values of impact isolation over the entire frequency range, more elaborate constructions such as B through D may be required.

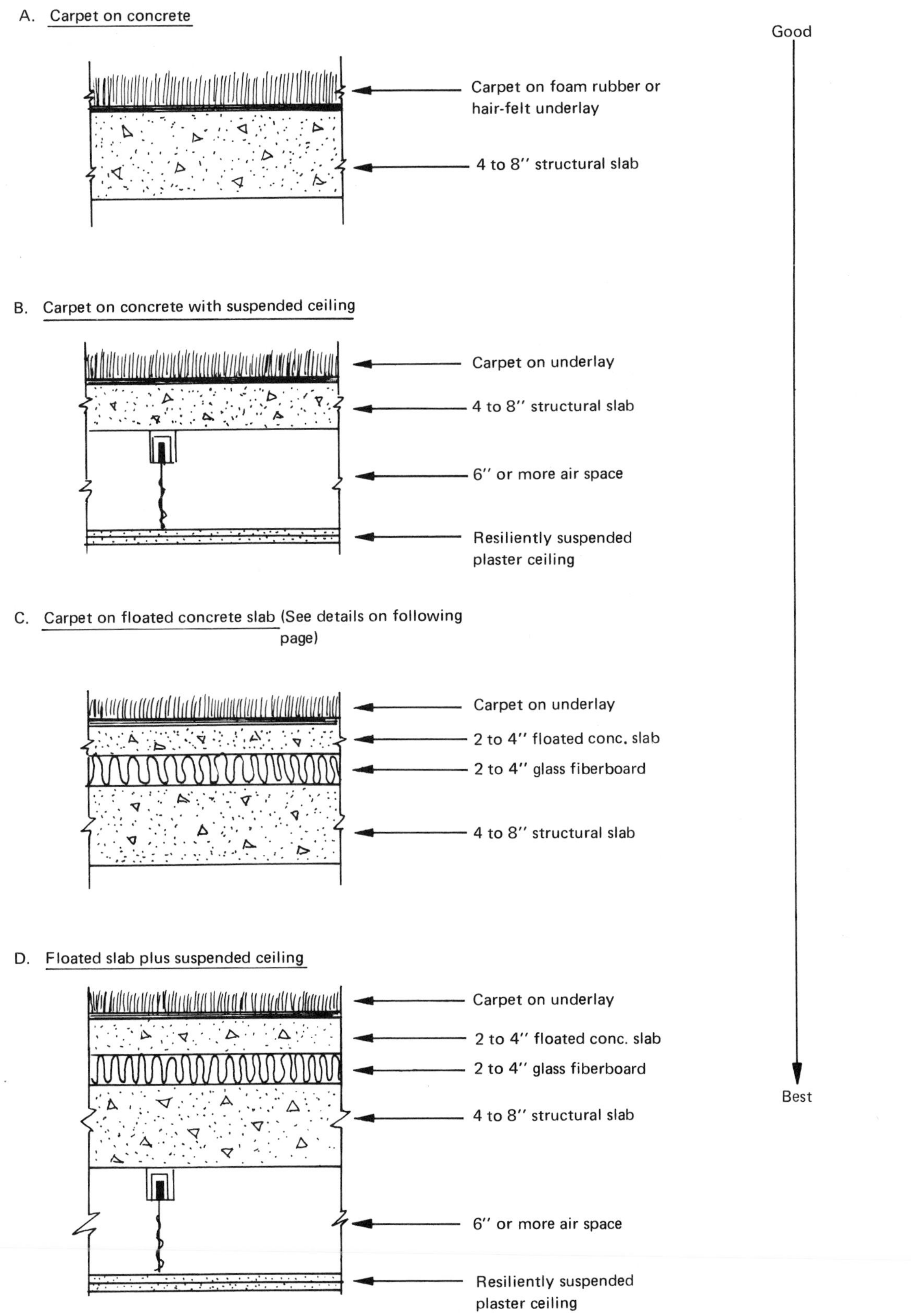

MECHANICAL SYSTEM NOISE AND VIBRATIONS: Typical Floated Floor Details

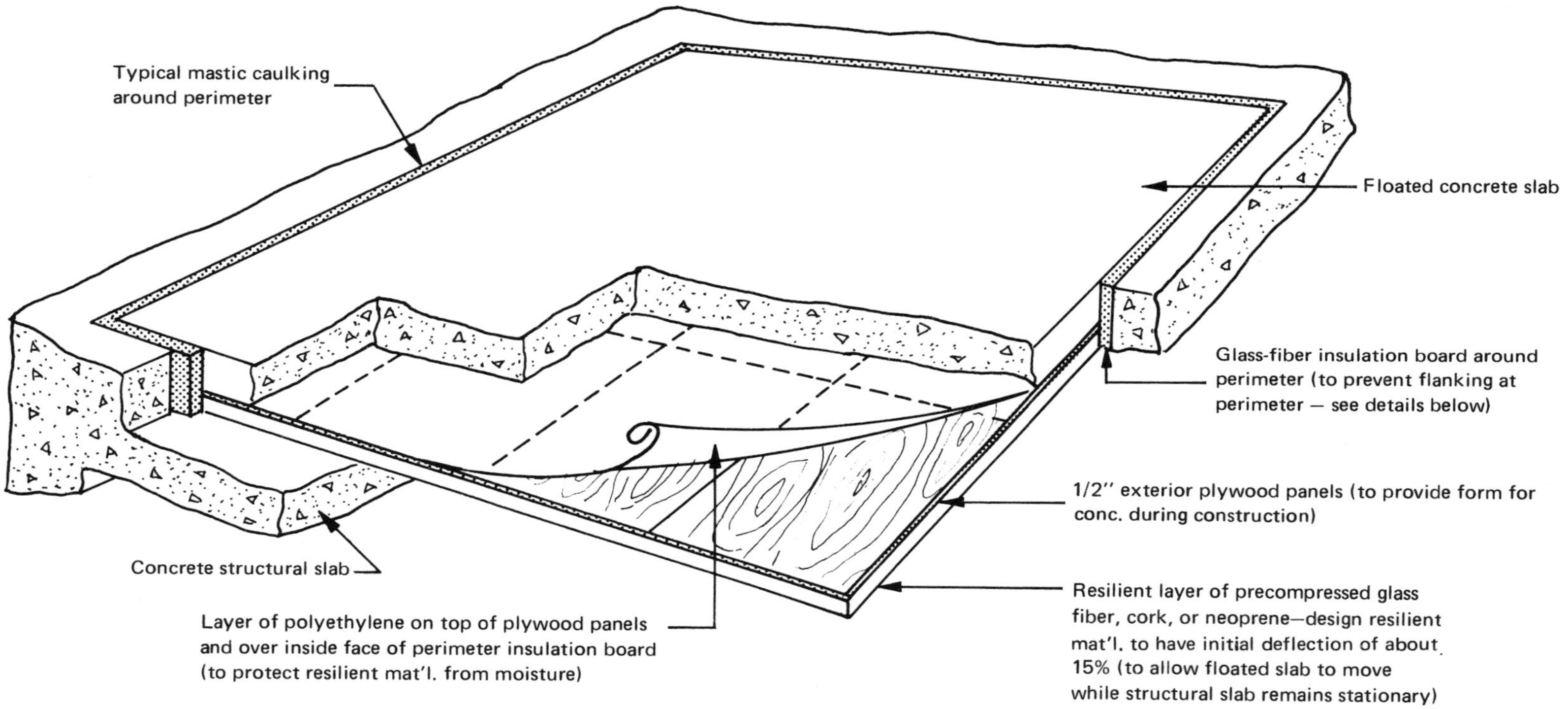

Partitions bordering floated floors should be designed as shown in A. If partitions are set on a floated floor, either a compressed resilient pad as in B or a shear in the floated surface as in C will provide a direct flanking path.

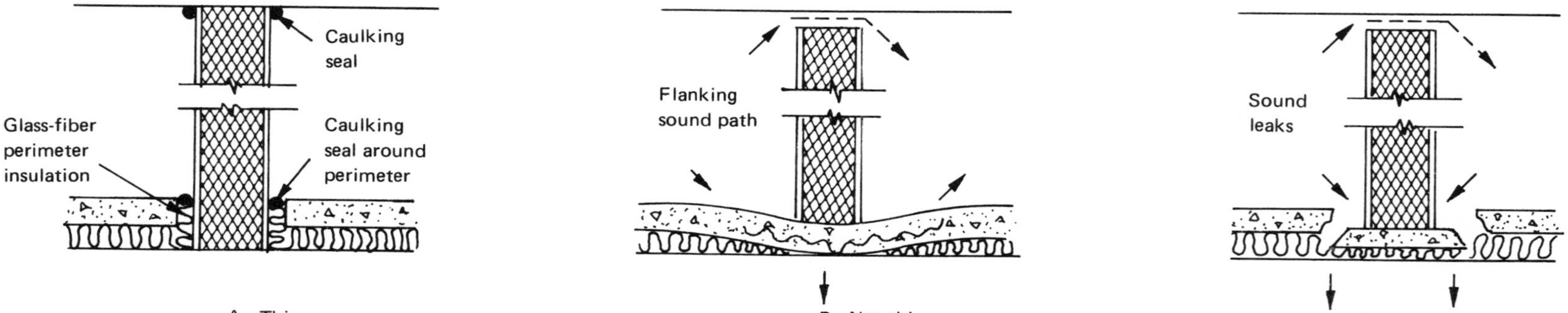

A. This

B. Not this

C. Not this

MECHANICAL SYSTEM NOISE AND VIBRATIONS: Impact Insulation Class (IIC) Overlay Contour and Grid

Trace IIC contour and grid shown below (without ordinate dB values) on a transparent overlay which can then be used to determine IIC ratings according to ISO procedures outlined on the following page.

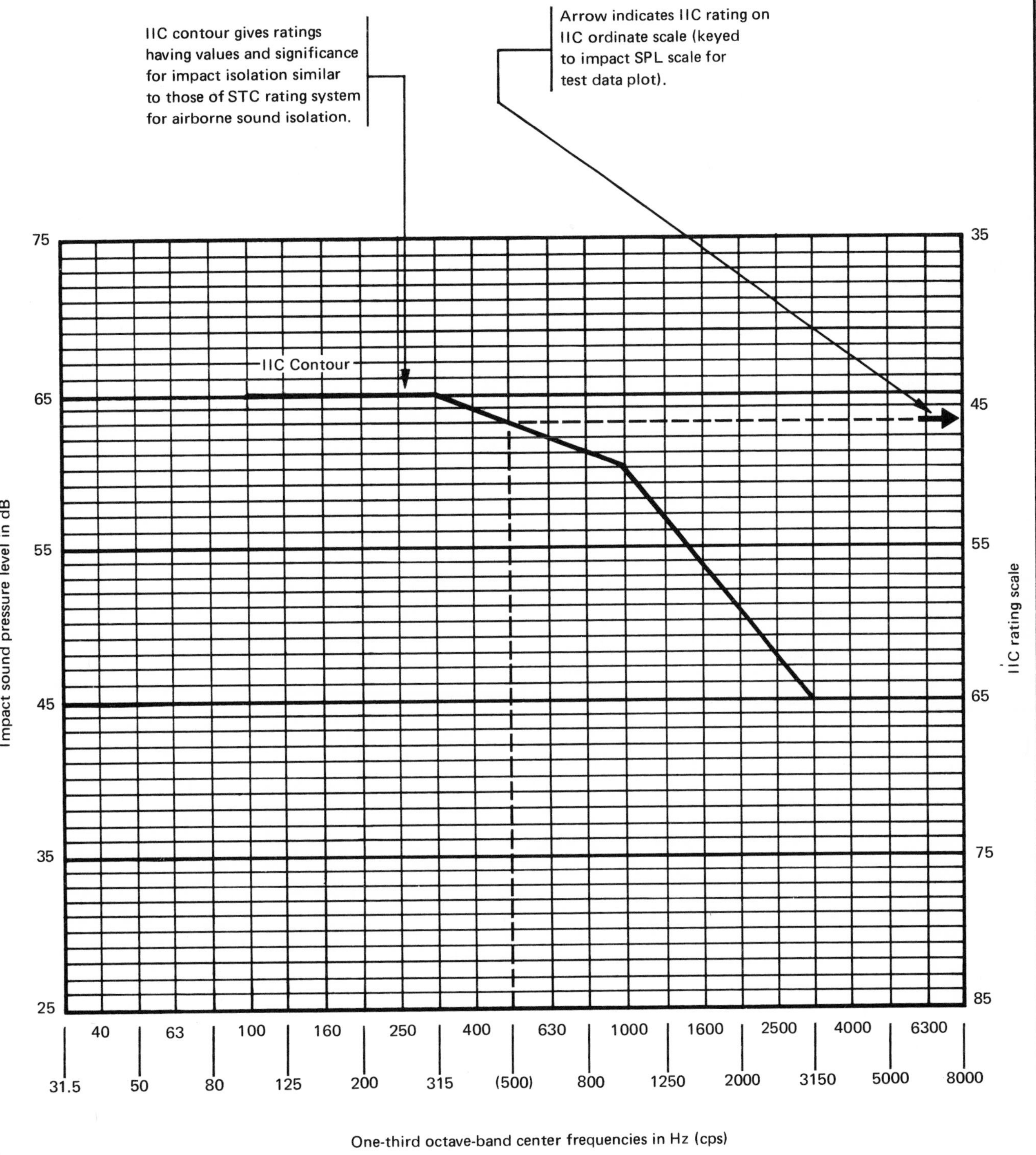

IMPACT INSULATION CLASS (IIC) RATING PROCEDURES

The IIC is a single-number rating of a floor-ceiling's impact sound transmission performance at different frequencies. The higher the IIC rating, the more efficient the floor-ceiling for reducing impact sound transmission in the test frequency range.

The IIC rating method is recommended by the Federal Housing Administration (FHA) as an improved rating method to supersede their previously recommended impact noise rating (INR) system. Like the INR, the IIC method is based on measurements of the absolute sound pressure levels produced in a room directly below the test floor on which the ISO standard tapping machine is operating. In this situation, lower transmitted impact sound pressure levels indicate better isolating performance. On the average, the IIC rating is about 51 points higher than the INR rating for a given construction.

IIC test data are measured at 16 third-octave bands with center frequencies from 100 to 3150 Hz. The test data, normalized to a room absorption of 108 sabins, are plotted against frequency and compared with a reference IIC contour. The IIC rating is easily determined by using a transparent overlay on which the IIC contour is drawn, as shown on the preceding page. Before the overlay is used, a new vertical scale is written on the right-hand side of the test-curve graph, with values decreasing in the upward direction. The right- and left-hand scales coincide at 55 dB and have the same number of decibels per graph divisions. The IIC contour is shifted vertically relative to the test data curve to as low a position as possible according to the following conditions:

1. The maximum deviation of the test curve above the contour at any single test frequency shall not exceed 8 dB.
2. The sum of the deviations of all 16 frequencies of the test curve above the contour shall not exceed 32 dB – an average deviation of 2 dB.

When the IIC contour is adjusted to the lowest position that meets the above requirements, the IIC rating is read from the right-hand vertical scale of the test curve as the value corresponding to the intersection of the IIC contour and the 500 Hz ordinate. For example, the position of the contour on the preceding page would give an IIC 47 rating, as shown.

References

1. "Field and Laboratory Measurements for Airborne and Impact Sound Transmission," International Organization for Standardization (ISO) Recommendation R140, ISO/R140 – 1960 (E).
2. "Rating of Sound Insulation for Dwellings," ISO Recommendation R717, ISO/R717 – 1968. ISO reports are available through the American National Standards Institute, 1430 Broadway, New York, N. Y. 10018.
3. "Airborne, Impact and Structure Borne Noise Control in Multifamily Dwellings," U. S. Department of Housing and Urban Development FT/TS-24, January, 1968.

MECHANICAL SYSTEM NOISE AND VIBRATIONS: Impact Insulation Class (IIC) for Floor-ceilings in Dwelling Units

FHA impact criteria for grade II dwelling units (i.e., residential urban and suburban with "average" noise environment) can be found in the table below at the intersection of the desired overhead source room row and receiving room column. For example, an IIC 57 is required for a kitchen over a living room (be sure plumbing fixtures are properly vibration isolated).

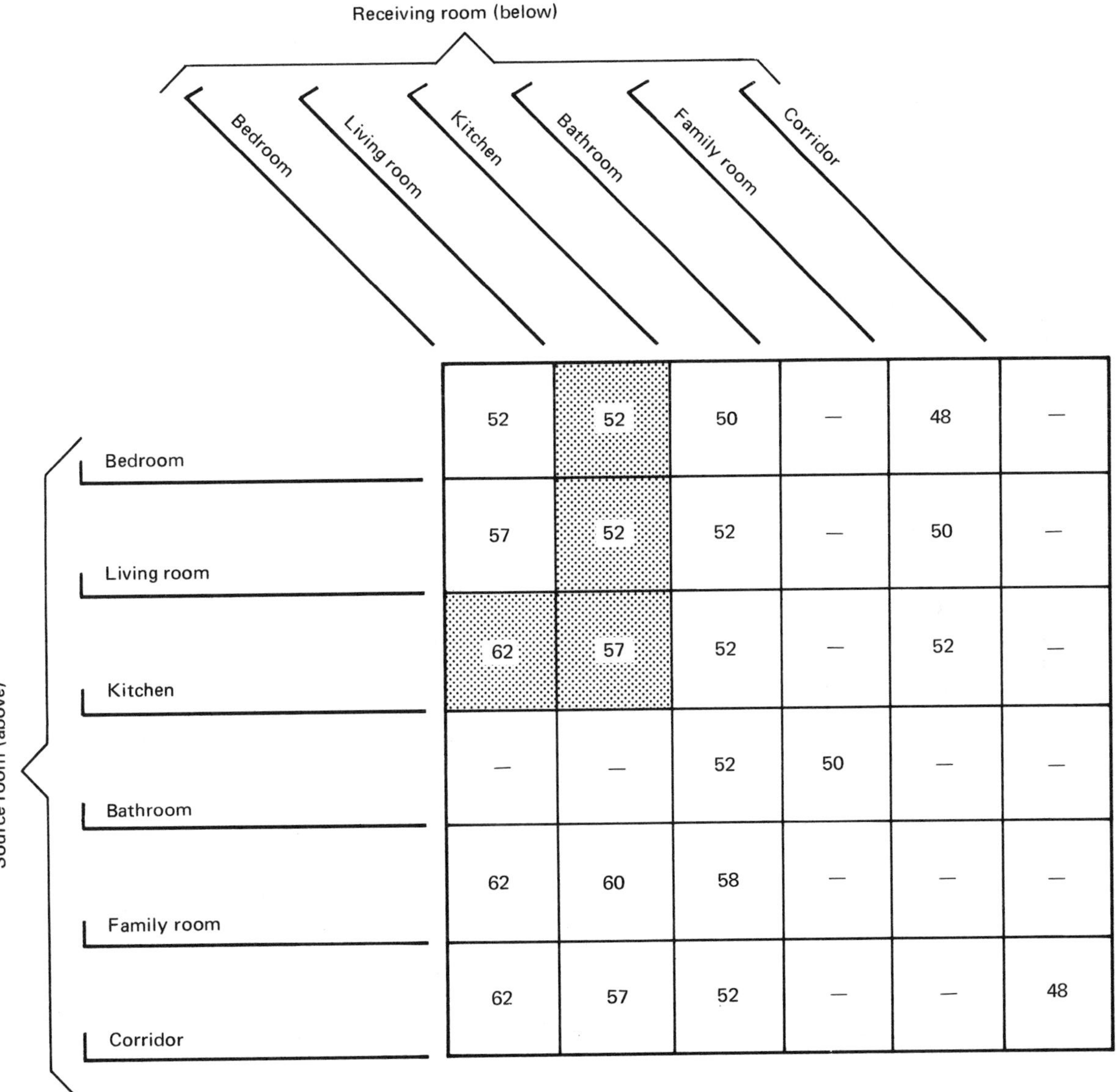

Source room (above) \ Receiving room (below)	Bedroom	Living room	Kitchen	Bathroom	Family room	Corridor
Bedroom	52	52	50	—	48	—
Living room	57	52	52	—	50	—
Kitchen	62	57	52	—	52	—
Bathroom	—	—	52	50	—	—
Family room	62	60	58	—	—	—
Corridor	62	57	52	—	—	48

NOTE: This table is also valid for grade I (suburban and peripheral suburban residential areas in "quiet" locations) and grade III (urban areas with "noisy" environments) dwelling units. For grade I add 3 to the table values, and for grade III subtract 4.

FLOOR-CEILING IIC AND STC IMPROVEMENTS

Modification to Basic Construction	Typical IIC Improvement	Typical STC Improvement
A. Wood-joist Systems:		
1. Resiliently suspended ceiling	8	10
2. Floated floor, i.e., flooring nailed to 1 1/2-by 2-in. wooden battens on glass fiberboard	8	10
3. Sound-absorbing material in air space (e.g., fuzz added to improvement A-1)	7	3
B. Concrete-slab Systems:		
1. Resiliently suspended ceiling	8	10
2. Floated floor – see A-2, above (Note greater improvement than obtained in wood-joist construction due to relative rigidity of concrete-slab base under the glass fiberboard.)	20	12
3. Sound-absorbing material in air space	5	3
4. Floated floor – concrete slab on 2-in. glass fiberboard	15	10
5. Wood flooring at least 1/2 in. thick, set in mastic	7	0
The floor-finish materials listed below provide improvements to both wood-joist and concrete-slab systems as follows:		
1. Vinyl tile	0	0
2. 3/32-in. linoleum	4	0
3. 1/4-in. cork tile	12	0
4. Carpet on foam rubber underlay	10–25	0

CHECKLIST FOR EFFECTIVE MECHANICAL SYSTEM NOISE CONTROL

- Avoid attaching vibrating or noisy equipment directly to the structural surfaces in a building.

- Layers of soft, resilient materials can be used under the wearing surface of floors to isolate the equipment room from the structure, or between the bases or supports of vibrating equipment and the structure to minimize transfer of vibrations into the structure.

- Prefabricated commercial resilient equipment mounts and bases are available and when properly selected provide resilient, stable support to vibrating equipment.

- Pipe and conduit connections to vibrating equipment should be vibration isolated for a considerable distance to interrupt the transmission path. All electrical conduits and pipe connections should be flexible and "floppy" when possible. Avoid metal-to-metal contact.

- Avoid constructions such as untreated mechanical ducts or rigid conduits, which can act as speaking tubes to transmit sound from one area to another. Line common ducts with glass fiber and, where they pass through walls, floors, etc., isolate them from the structure with resilient materials and caulk the perimeter airtight.

SECTION 6

Room Acoustics

xxx5

ROOM ACOUSTICS: Ancient Theaters

Some ancient, open-air Greek and Roman theaters had good listening conditions suitable for speech, drama, chorus, and recitals. The Greek theaters, for example, were usually located on steep hillsides in quite rural locations. Successful sites had few gusty winds, as air causes noise as it blows past the audience. Some characteristic design features are shown below. Today, however, open-air theaters should be designed only according to up-to-date concepts. Also, to achieve reverberation, necessary for modern music, drama, and opera (especially since the nineteenth century), an enclosure is required.

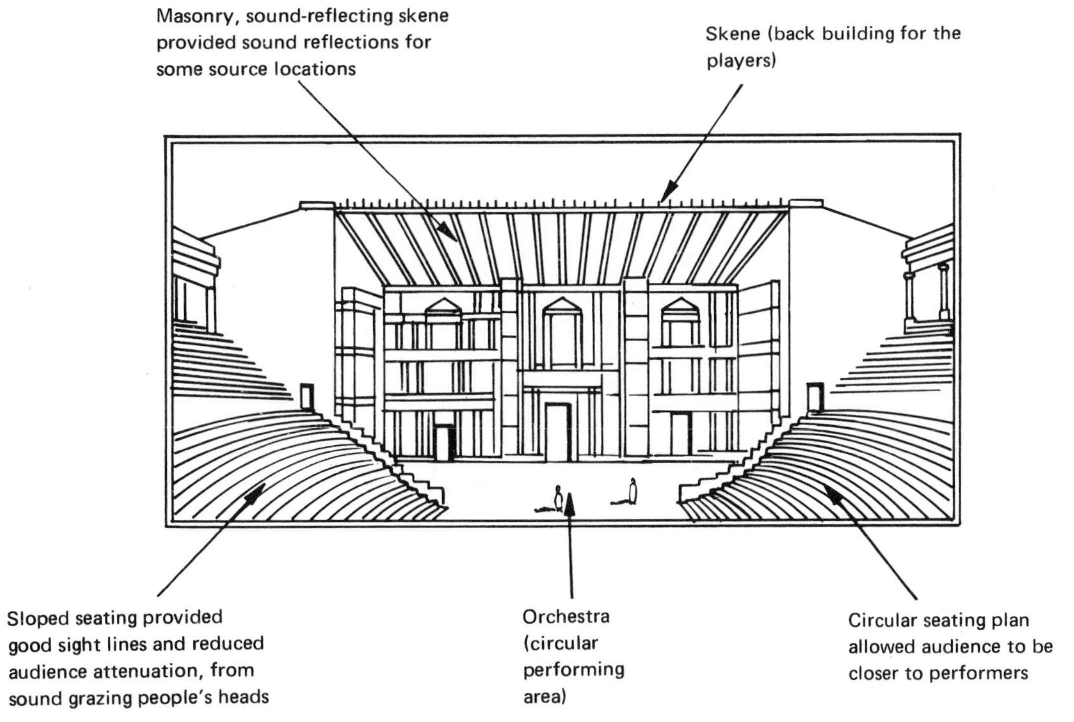
Masonry, sound-reflecting skene provided sound reflections for some source locations
Skene (back building for the players)
Sloped seating provided good sight lines and reduced audience attenuation, from sound grazing people's heads
Orchestra (circular performing area)
Circular seating plan allowed audience to be closer to performers

NOTE: **For a comprehensive study of the development of the ancient theater, see M. Bieber, The History of Greek and Roman Theater, Princeton Univ. Press, Princeton, N. J., 1961.**

ROOM ACOUSTICS: Audience Seating

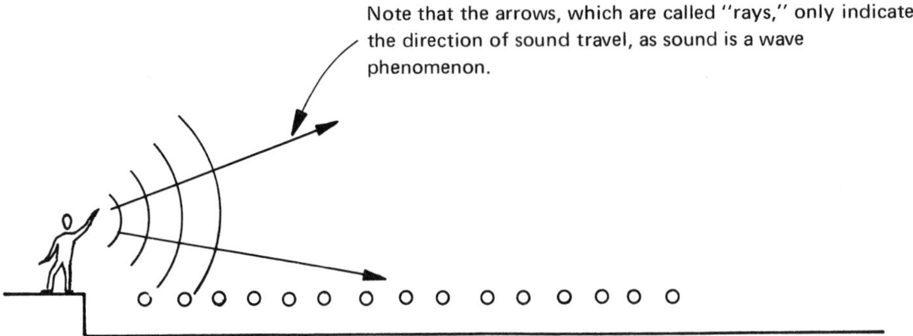

A. Sound level outdoors falls off with distance and from audience attenuation.
Sound spreads outward, losing energy according to the inverse-square law and also by audience attenuation, as sound grazes the seated audience.

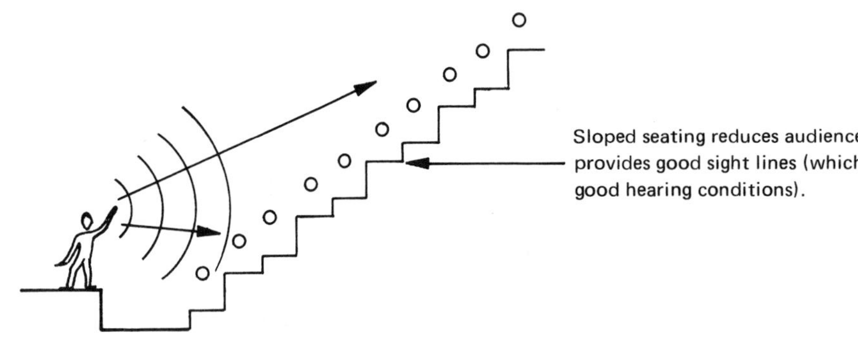

Sloped seating reduces audience attenuation, and provides good sight lines (which usually mean good hearing conditions).

B. Sound level outdoors falls off only with distance.

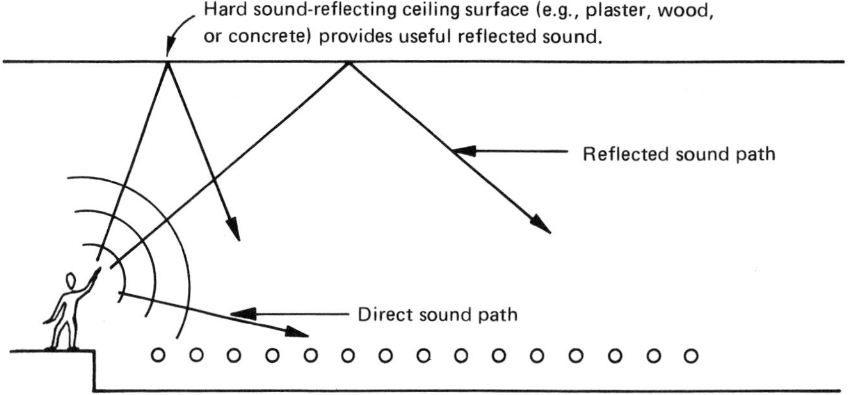

C. Reflected sound from ceiling reinforces direct sound level indoors.

Note also that a hard, sound-reflecting enclosure (e.g., band shell) near the source can greatly improve listening conditions outdoors!

ROOM ACOUSTICS: Reflection, Diffusion, and Diffraction

- Reflection

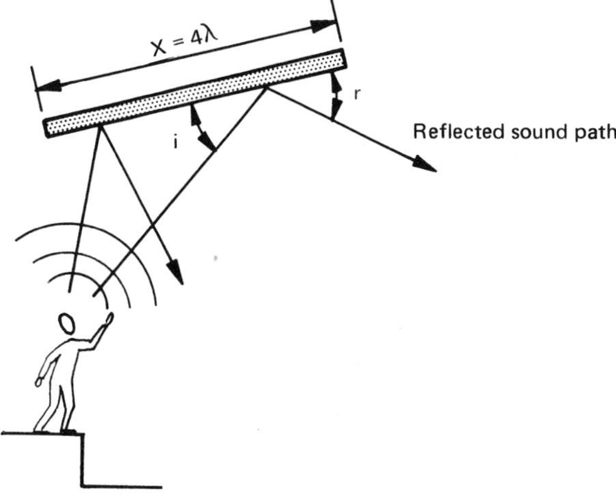

X ≥ 4λ

Reflection is the return of a sound wave from a surface. If the surface is large compared with the wavelength of sound, the angle of incidence "i" will equal the angle of reflection "r." For example, 1000 Hz corresponds to a wavelength of 1 ft so surface dimensions (length & width) of about 4λ or 4 ft will reflect sound at 1000 Hz and above.

- Diffusion

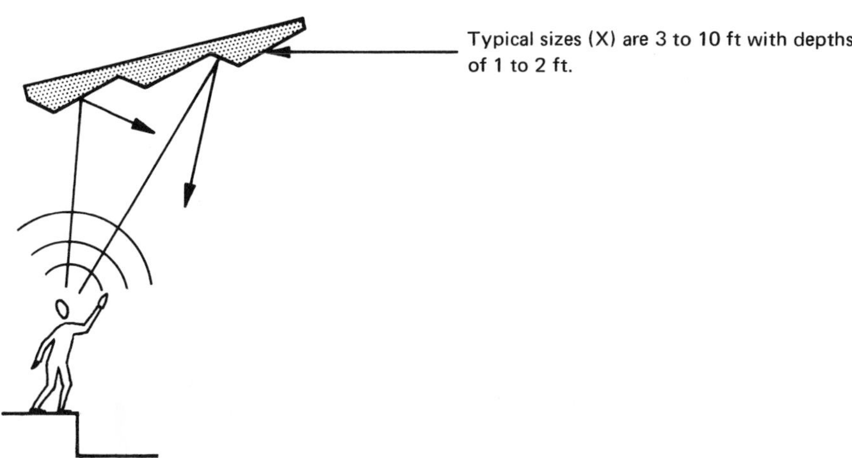

Typical sizes (X) are 3 to 10 ft with depths of 1 to 2 ft.

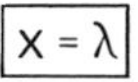

Diffusion is the scattering or random distribution of a sound wave from a surface. It occurs when the surface size equals the wavelength of sound. Diffusion does not "break up" sound—sound is not fragile or brittle! It merely changes its direction when it strikes a hard-surfaced material.

- Diffraction

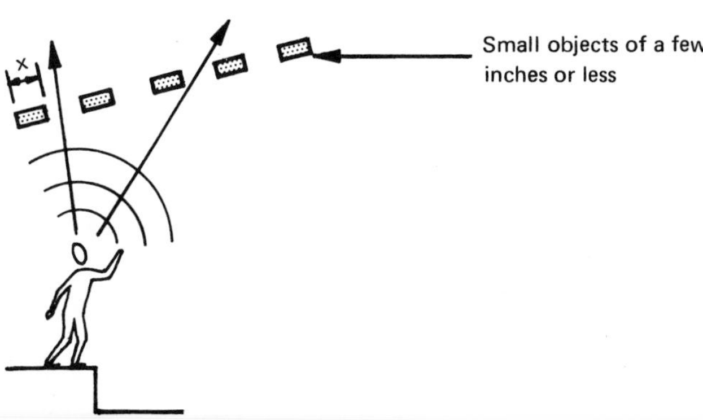

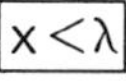

Diffraction is the bending or "flowing" of a sound wave around an object (or through an opening). For example, a car concealed by a building can be heard even when it cannot be seen—the sound bends around the corner!

ROOM ACOUSTICS: Correlation of a Sound's Wavelength in Air and Its Frequency under Normal Conditions (Temp., Atmos. Press., etc.)

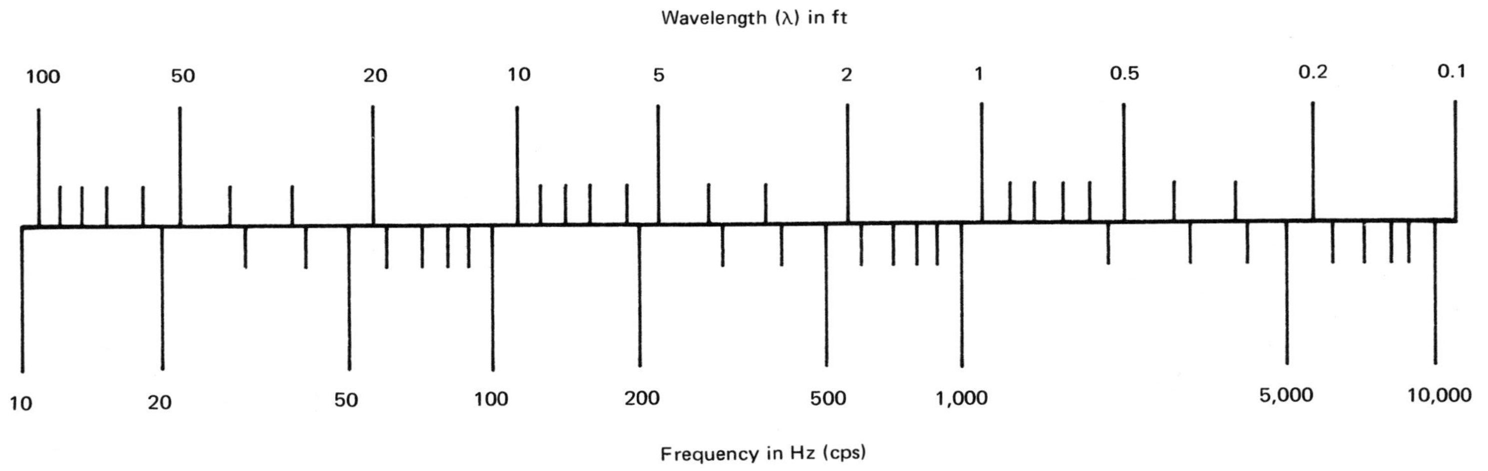

Wavelength (λ) in ft

100 50 20 10 5 2 1 0.5 0.2 0.1

10 20 50 100 200 500 1,000 5,000 10,000

Frequency in Hz (cps)

ROOM ACOUSTICS: Distribution of Reflected Sound

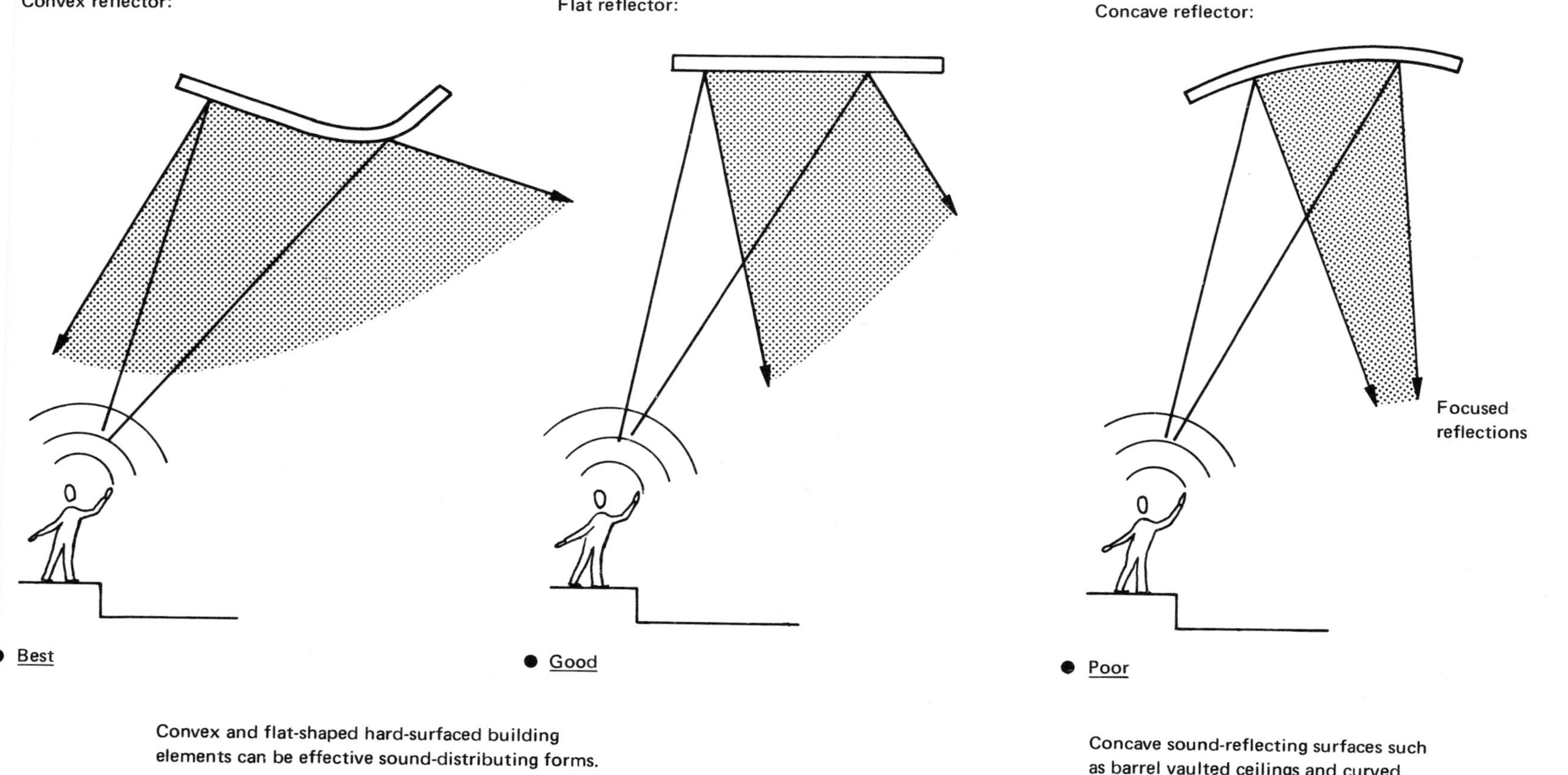

- Best
- Good
- Poor

Convex and flat-shaped hard-surfaced building elements can be effective sound-distributing forms.

Concave sound-reflecting surfaces such as barrel vaulted ceilings and curved auditorium rear walls can focus sound, causing echoes. Because concave surfaces focus sound, they are also poor distributors of sound energy and therefore should be avoided where sound reflections are important.

NOTE: **Shading indicates area of distribution between rays shown.**

RAY-DIAGRAM ANALYSIS

Ray-diagram analysis is used to (1) study the effect of room shaping on *sound distribution* and (2) identify possible *echo*-producing surfaces. A ray diagram is an acoustical analogy to the regular or specular reflection of light where the angle of incidence i of an impinging sound wave equals the angle of reflection r. That is, it assumes that sound waves are reflected from surfaces in the same way a billiard ball, without spin, rebounds from a cushion. The following page shows basic ray-diagram graphics. Some limitations in ray-diagram studies are as follows:

1. Sound reflects in the manner indicated by ray-diagramming only when surface dimensions are large relative to the wavelength of sound being considered.
2. Normally the sources of speech or music will not radiate from a fixed position. Optimum room shaping therefore depends on a careful balance of the best sound distribution from several source positions to the listening area.
3. A detailed evaluation of diffusion or scattering of sound from room surfaces is not possible with ray-diagrams. Accordingly, models allowing frequency-scaled acoustical studies are often used in design, particularly in rooms where music listening is important.

In spite of these limitations, ray-diagramming is still an important design tool in establishing optimum room shape. The table below is a design guide that can be used with ray-diagram analysis to check general listening conditions. The difference in length between the reflected sound path and the direct sound path at any listening position is directly related to the time difference which the ear detects. For example, if sound reaches the listener's ear later than about 1/17 sec after the direct sound, it can be heard as an echo. A familiar example of this would be the echoes from remote cliffs in mountain regions, where one can experience reflected sounds that are distinct repetitions of the direct or original sound. The sound path difference in feet from a sound delayed by 1/17 sec (i.e., about 0.06 sec) can be found as follows: distance = velocity × time = 1130 × 0.06 = 68 ft.

Sound Path Difference, ft	*Listening Conditions*
Less than 28	Excellent for speech and music
28–40	Good for speech, fair for music
40–50	Marginal
50–68	Negative
Greater than 68	Echo if strong enough

References

1. V. O. Knudsen: Architectural Acoustics, *Sci. Amer.*, Vol. 209, no. 5, November, 1963.
2. R. L. McKay: "Notes on Architectural Acoustics," Stipes Publishing, Champaign, Ill., 1964.

ROOM ACOUSTICS: Ray-diagram Graphics

An inexpensive protractor for measuring angles, pencil, and paper are all the equipment required for ray-diagram calculations. Shown below is an auditorium section with path differences calculated to a front and middle-rear audience location from a typical source location.

Symbols showing source to listener sound paths:

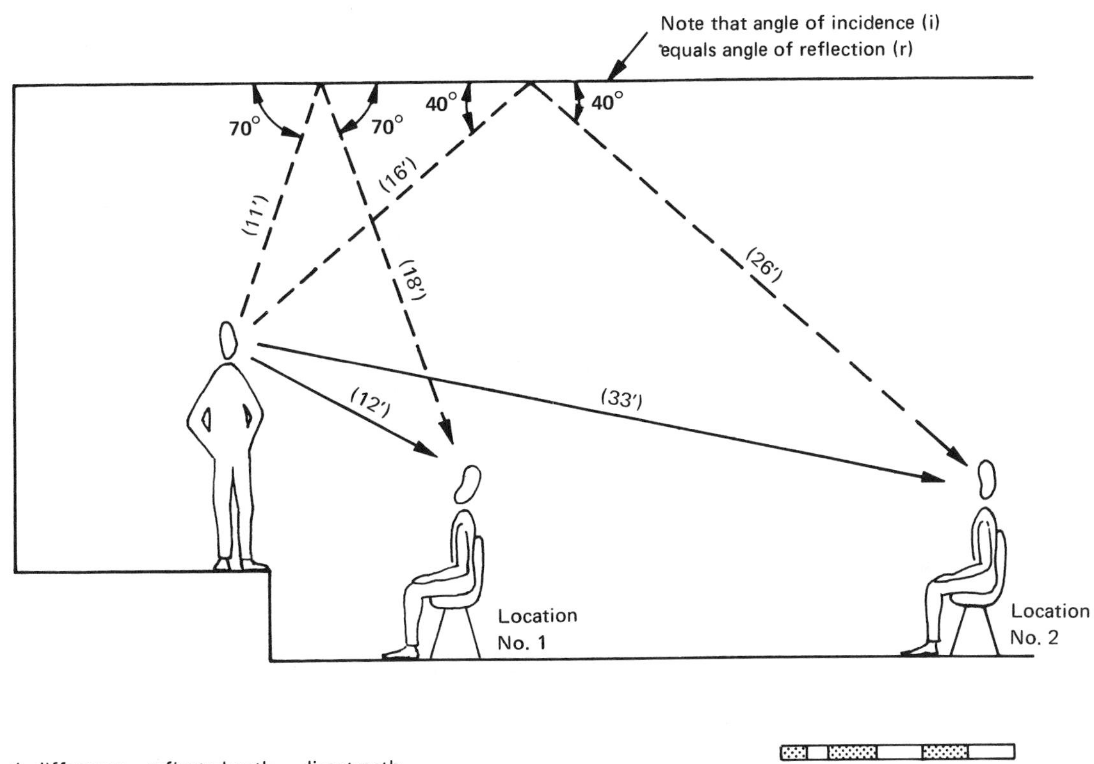

Path difference = reflected path − direct path

0 2 4 6 8 10 ft

EXAMPLES – RAY-DIAGRAM MEASUREMENTS **(distances shown in parentheses on above section)**

A. Front location no 1:

Path difference = (11 + 18) – (12) = 17 ft < 28 ft

Excellent for speech and music

B. Middle-rear location no 2:

Path difference = (16 + 26) – (33) = 9 ft < 28 ft

Excellent for speech and music

ROOM ACOUSTICS: Sound Paths in Auditoriums

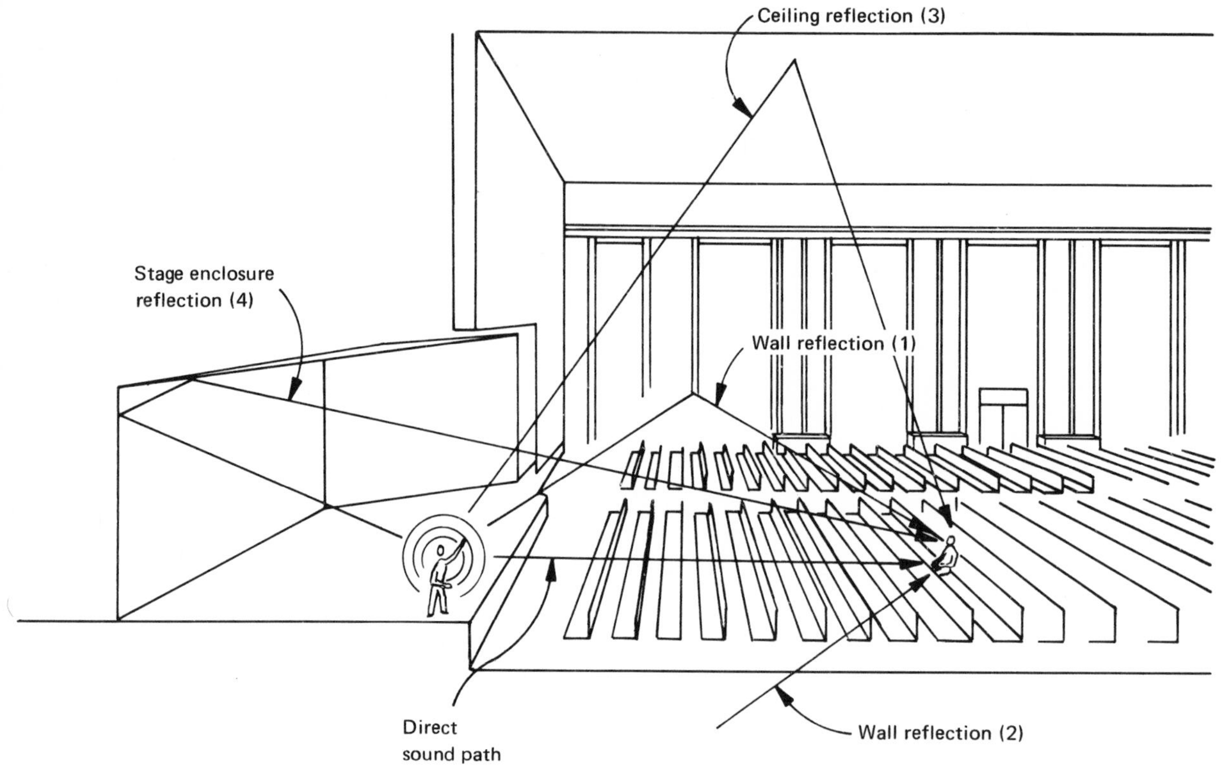

Direct and reflected sound paths

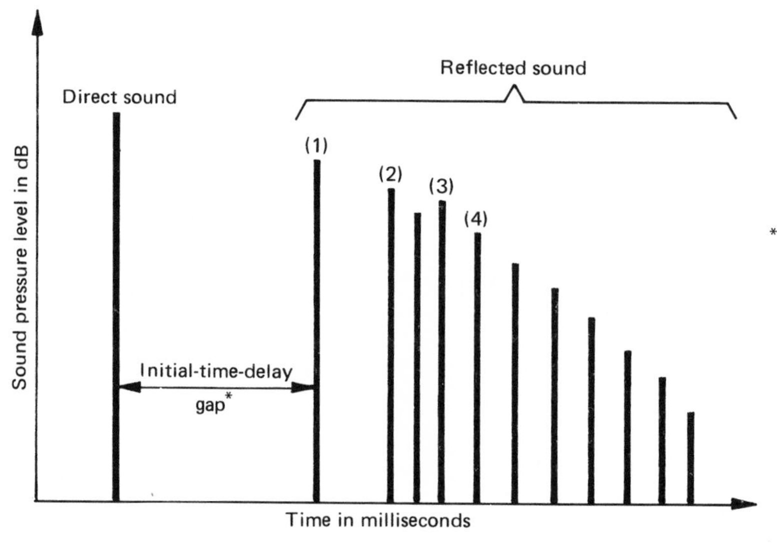

* The initial-time-delay gap should be less than 30 milliseconds (path diff. ⩽ 34 ft) for good listening conditions.

Time graph

The listener in the above auditorium will hear the direct sound first and, after the initial-time-delay gap, reflections from the walls, ceiling, stage enclosure, etc. This is indicated by the bars on the time graph, which show the direct sound followed by the reflected sounds (at lower sound pressure levels) arriving after an initial-time-delay gap.

ROOM ACOUSTICS: Ceiling Shape

Ray-diagram analysis indicates that the hard, sound-reflecting flat ceiling shown below provides useful sound reflections covering the entire seating area. However, by carefully reorienting the ceiling, as shown by the lower sketch, the extent of useful ceiling reflections can be increased so that the middle-rear seats actually receive reflections from both ceiling planes.

Flat ceiling

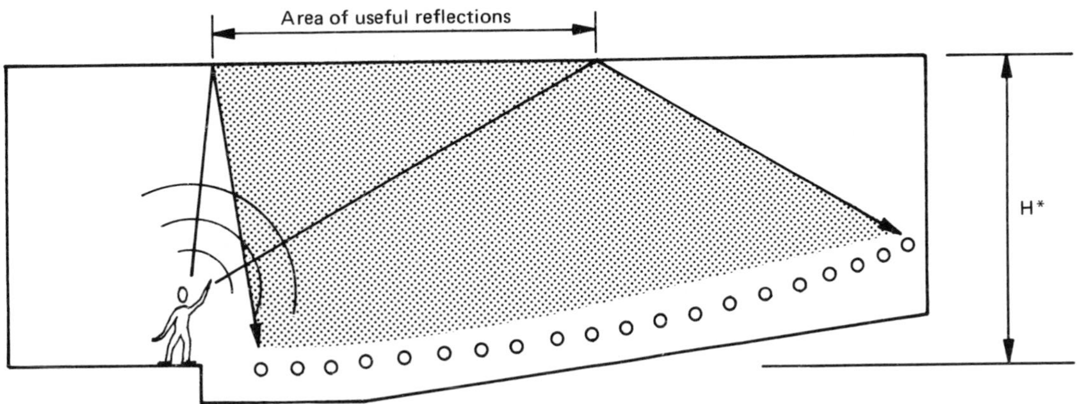

* Average ceiling height (H) in auditoriums w/upholstered seats and absorptive rear walls for echo control is approx. related to the mid-frequency reverberation time as follows:

$$H \simeq 20T$$

Reoriented ceiling

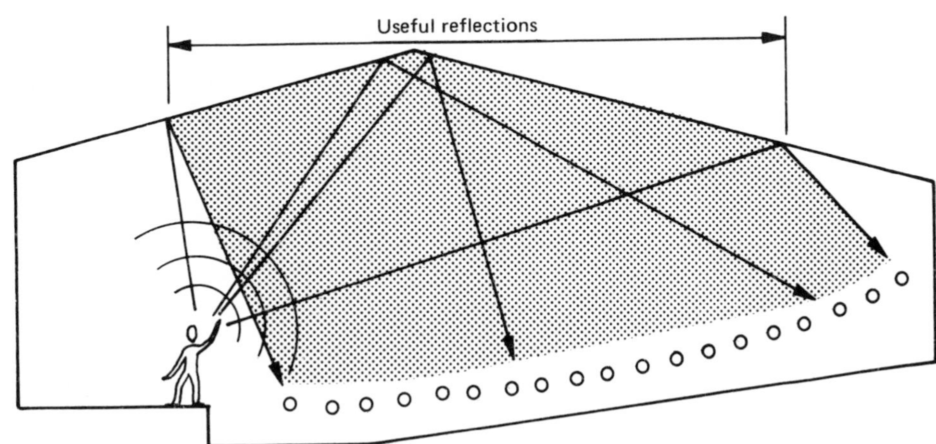

ROOM ACOUSTICS: Echo Control

Potential echo-producing surfaces should be treated with efficient sound-absorbing materials or shaped as shown below, where the forward ceiling is lowered and reoriented to provide useful reflections.

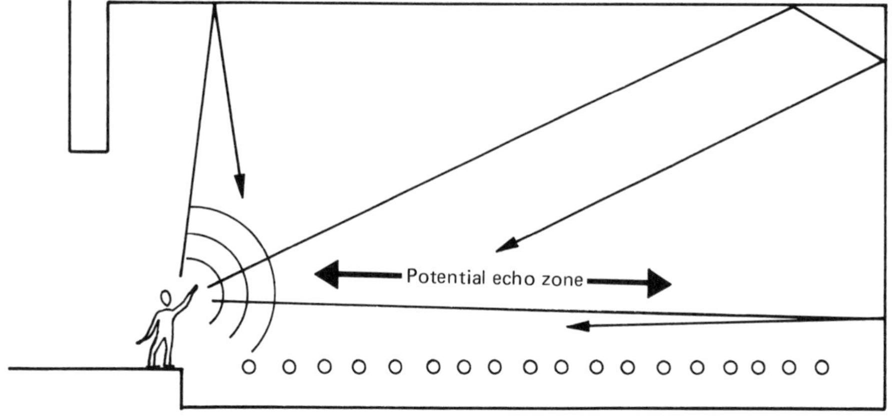

Potential echo-producing surfaces

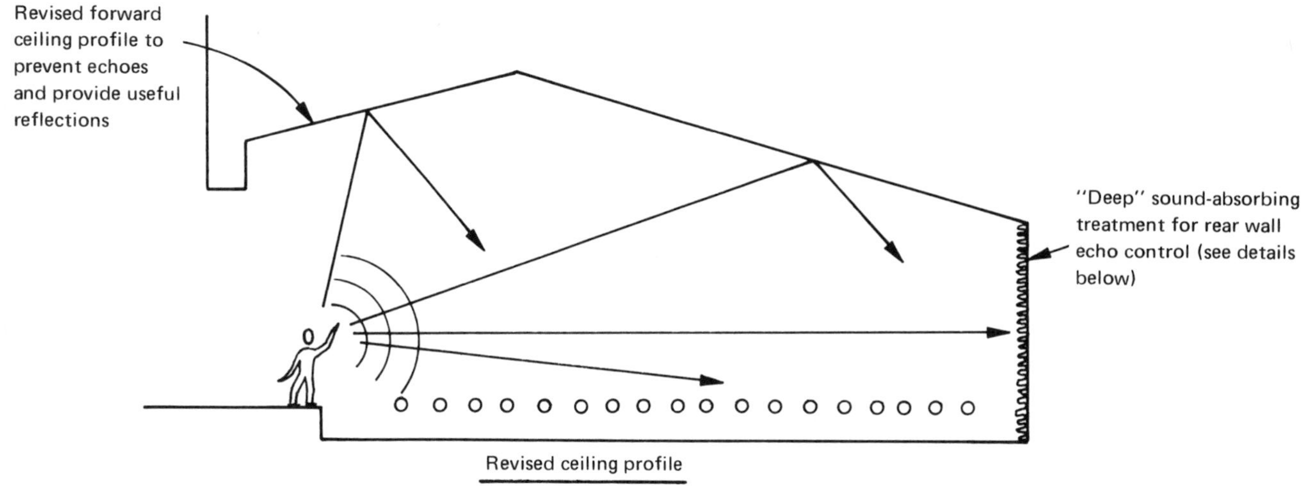

Revised ceiling profile

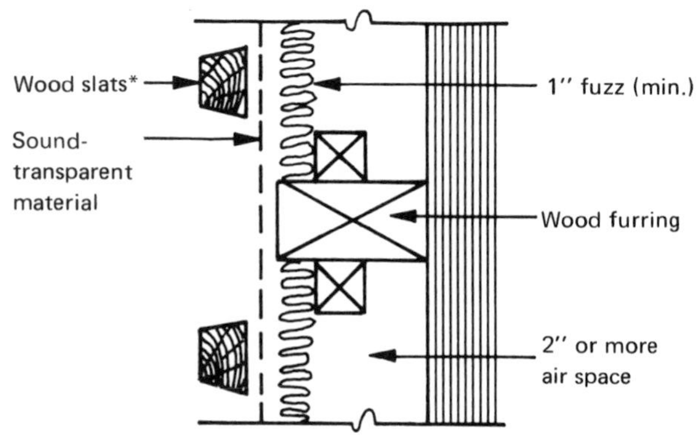

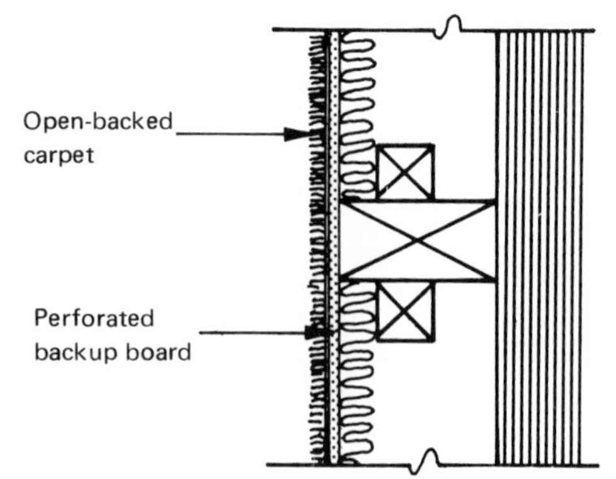

*At least 40% open

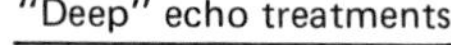

"Deep" echo treatments

ROOM ACOUSTICS: Rear Wall Echo Control Treatments

Problem

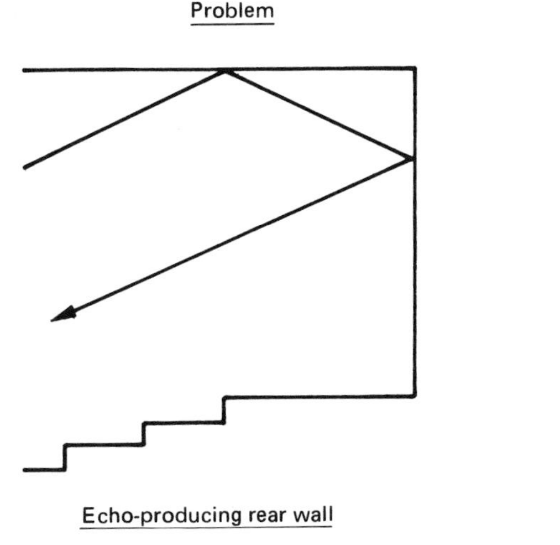

Echo-producing rear wall

Solutions

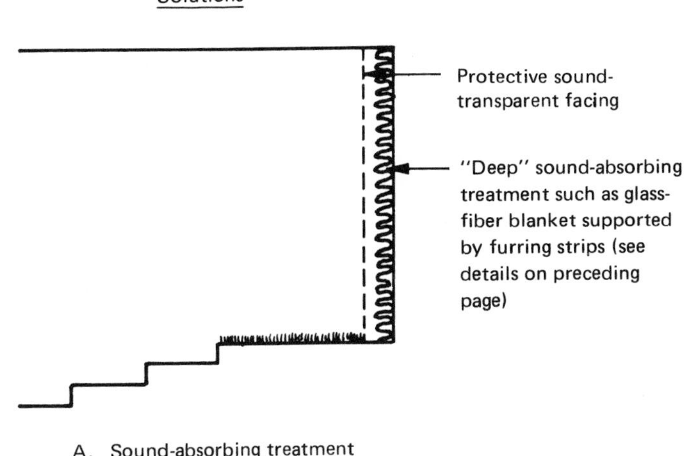

A. Sound-absorbing treatment

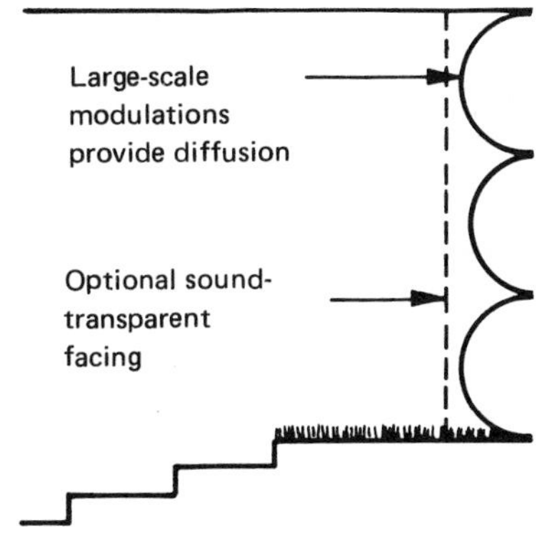

B. Surface modulations

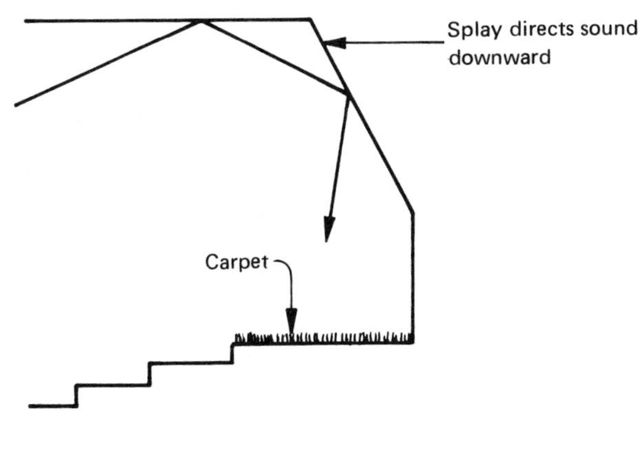

C. Splayed wall

ROOM ACOUSTICS: Side Wall Shape

Ray-diagram analysis is also important in the horizontal plane for study of side wall sound reflections. These reflections are very important in creating a subjective impression of space.

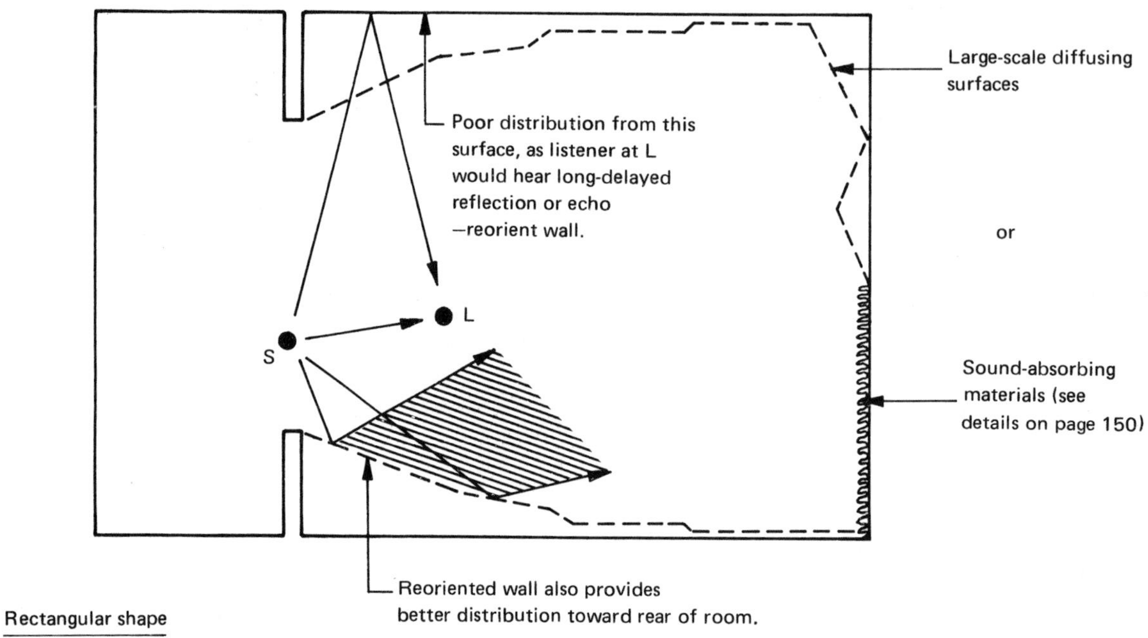
Large-scale diffusing surfaces
Poor distribution from this surface, as listener at L would hear long-delayed reflection or echo —reorient wall.
or
L
S
Sound-absorbing materials (see details on page 150)
Reoriented wall also provides better distribution toward rear of room.

Rectangular shape

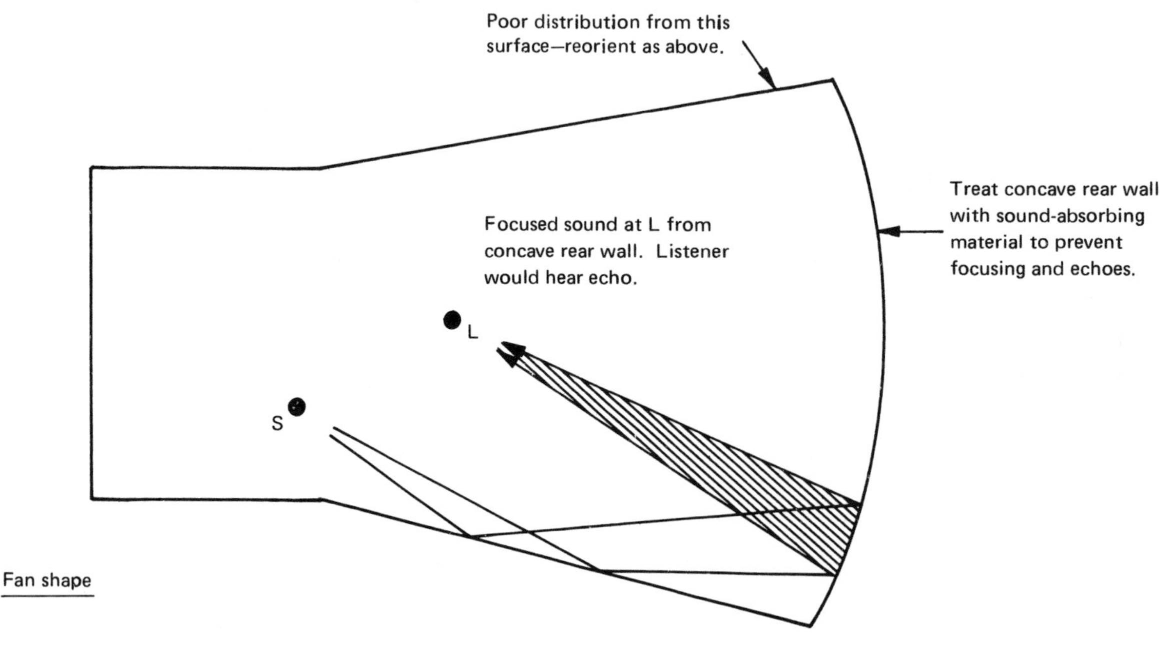
Poor distribution from this surface—reorient as above.
Treat concave rear wall with sound-absorbing material to prevent focusing and echoes.
Focused sound at L from concave rear wall. Listener would hear echo.
L
S

Fan shape

ROOM ACOUSTICS: Reverberation and Echo Patterns

Examples for the decay of sound in both large and small rooms are shown below. This kind of information is often important for the complete understanding of the acoustical quality of listening spaces.

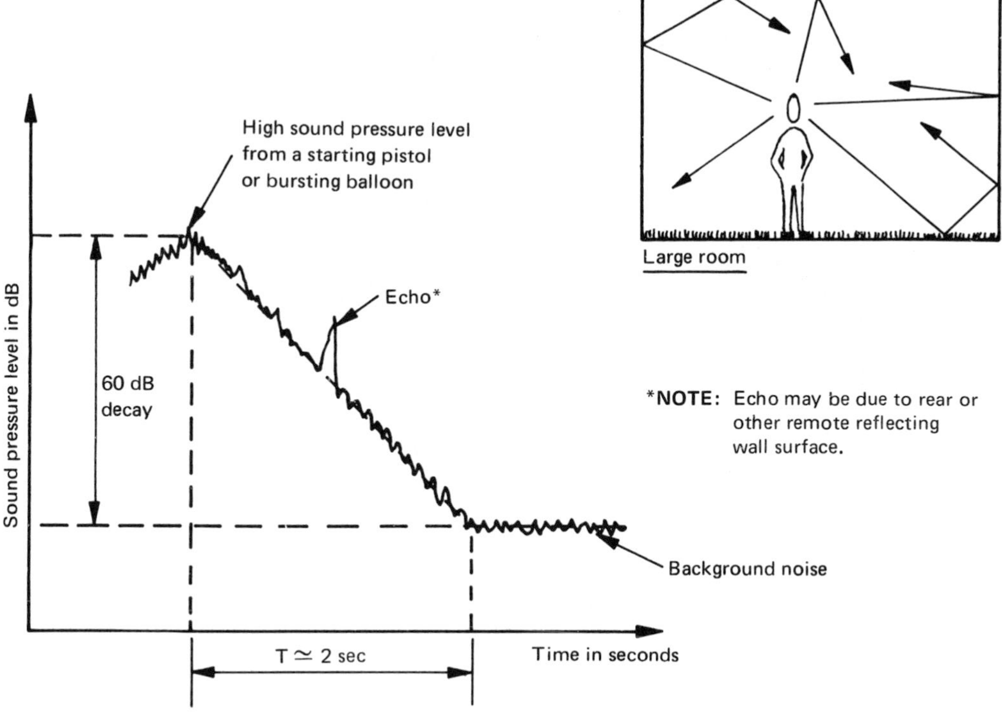

Reverberant decay of sound in a large room

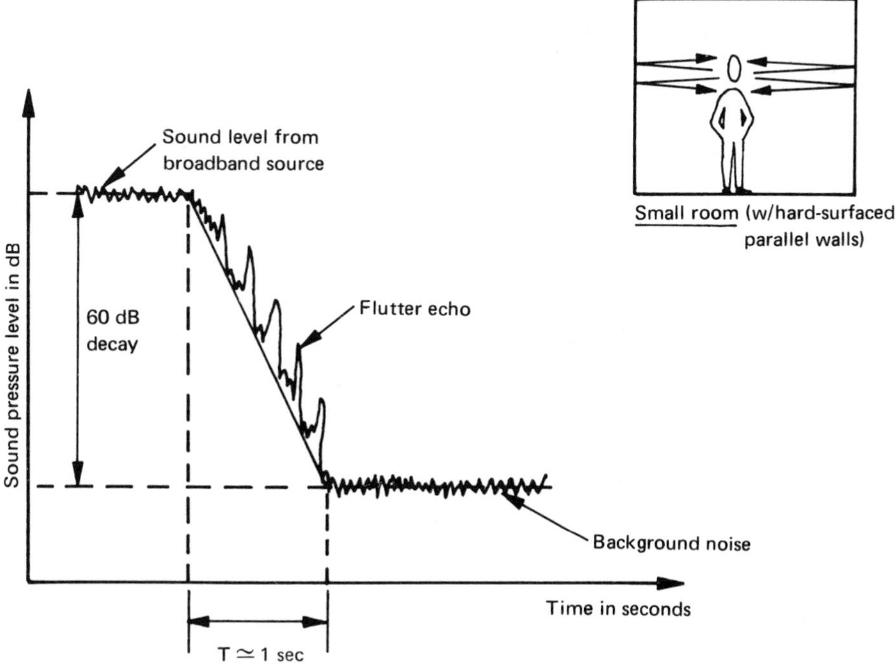

Reverberant decay of sound in a small room

NOTE: An articulation index (AI) can be derived from laboratory analysis of electronically recorded reverberation. For a discussion of this method, see J. P. A. Lochner and J. F. Burger, The Influence of Reflections on Auditorium Acoustics, J. Sound Vib., Vol. 4, pp. 426–454, 1964.

ROOM ACOUSTICS: Flutter Echo

Flutter echo is usually caused by the interreflection of sound between opposing parallel or concave surfaces. Flutter is normally heard as a high frequency "ringing" or "buzzing." Flutter can be prevented by (1) reshaping to avoid parallel surfaces, (2) providing sound-absorbing treatment, or (3) surface breakup with splayed elements. A 1:10 splay (or tilt) of one of the walls will normally provide sufficient flutter control.

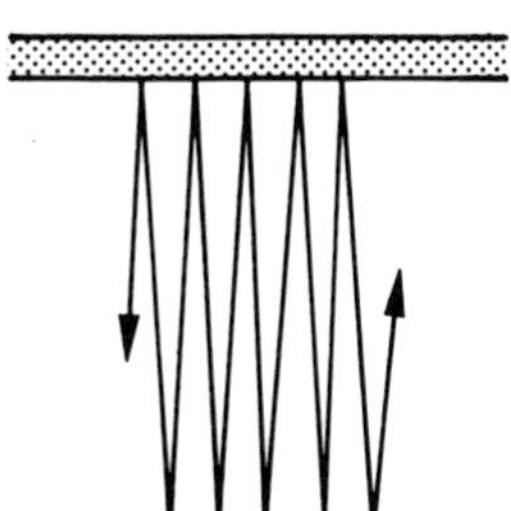

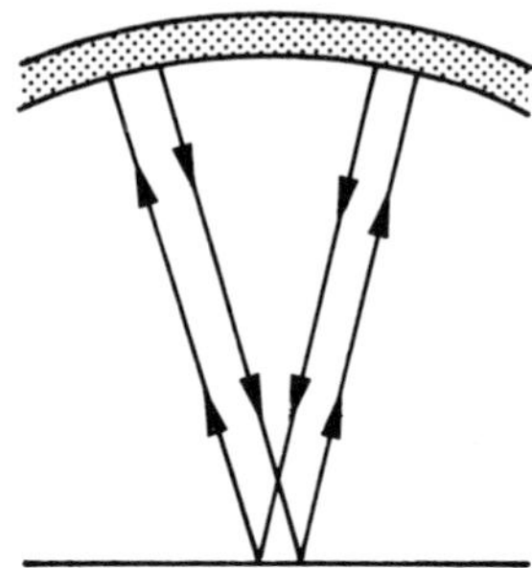

Flutter echo conditions

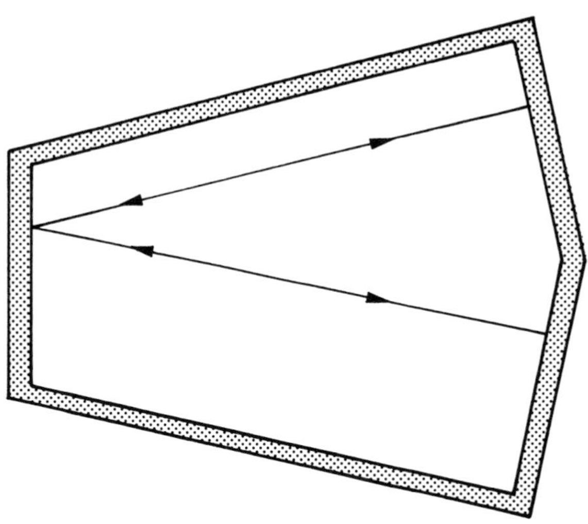

Flutter in room with nonparallel walls

The condition shown above is called "pitched-roof flutter."

ROOM ACOUSTICS: Small Rooms

Small rooms can resonate at certain audible frequencies; this is called the "bathroom tenor" effect. Specific preferred room dimension ratios are often quoted to provide even distribution of low frequency sound energy. However, since most rooms contain absorption from carpets, curtains, etc., the choice of exact dimension ratios is usually unimportant. In studios for music, sufficient wall absorption can be provided, as shown below. By placing sound-absorbing treatment on adjacent walls, the portion of the sound energy confined to reflective surfaces will be minimized.

A. Sound-absorbing surfaces opposite sound-reflecting surfaces

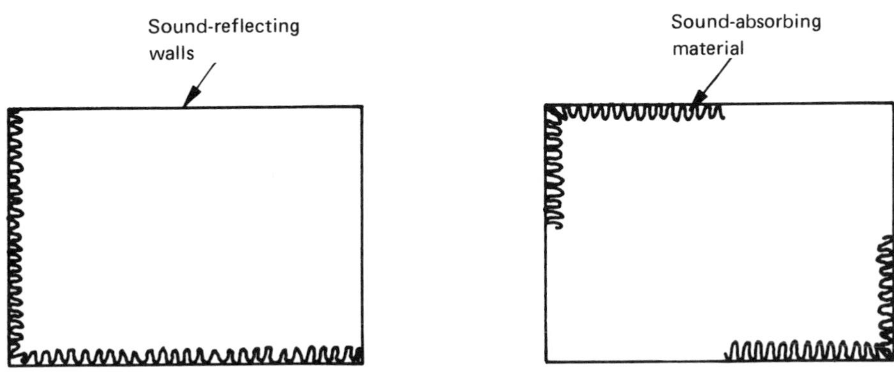

Sound-absorbing treatment should be furred out from backup wall, or use heavy fabric curtains hung 100% "full," which means 2 ft of curtain to 1 ft of surface width.

B. Nonparallel wall surfaces

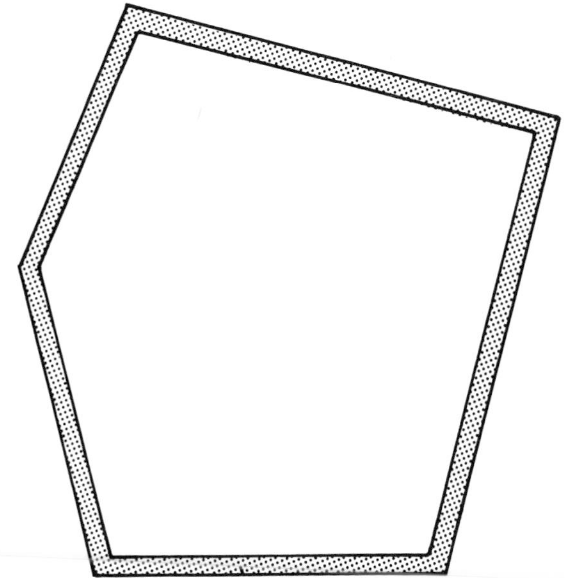

NOTE: Sound-absorbing ceiling or fully carpeted floor will be required if walls are hard-surfaced.

AREA EFFECT

The efficiency of a sound-absorbing material can be affected by its location and distribution in a room. For example, 25 panels of sound-absorbing material, each 2 by 2 ft, spaced in a checkerboard pattern on a 200 sq ft plaster ceiling, will absorb more sound per panel than uniform treatment with 50 panels of the same material. This increase in efficiency is called the "area effect."

The area effect is due to sound diffraction around the perimeter of the spaced absorbing panels and to the additional absorption provided by the exposed panel edges. Sound energy reflected from the hard-surfaced plaster immediately around the absorbent edges in this example tends to spill over onto the absorbing panels. Therefore, in effect, the spaced absorbing material actually absorbs more sound than would be accounted for by its physical area. Although the efficiency per unit is greater for spaced absorbers, the total room absorption will usually be less than a uniform ceiling coverage with the same material because of the smaller number of panels used.

Several manufacturers make sound-absorbing units specifically for distributed applications. The data are normally presented in terms of sabins per unit at the recommended spacings. Where a continuous application of conventional materials is not feasible, e.g., in industrial plants with ceilings remote from the source of noise, suspended spaced absorbers are particularly useful. Typical shapes are cones, flat baffles, prisms, parallelepipeds, tetrahedrons, etc. Data for suspended flat baffles are listed on page 35.

ROOM ACOUSTICS: Curved Forms

Cylindrical forms can cause annoying focusing problems, which can be corrected by treatments A and B shown below.

Problems

Solutions

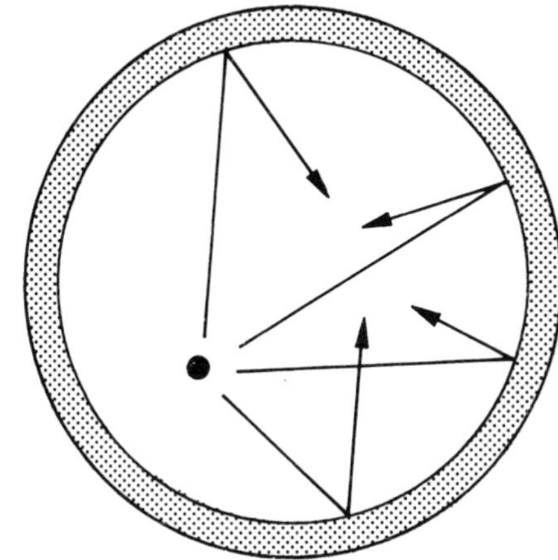

Focused sound

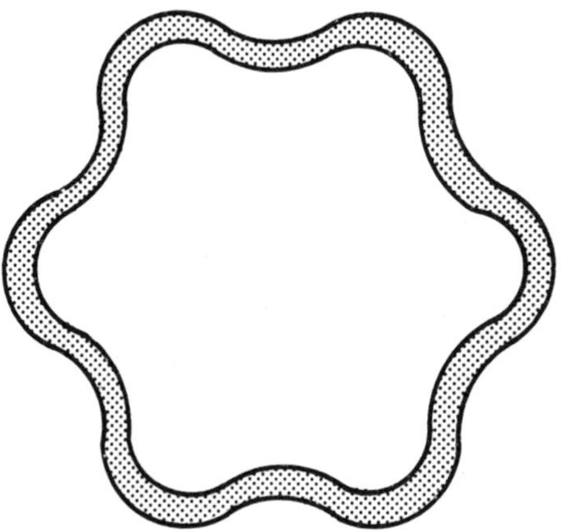

A. Surface undulations

Large-scale, random-sized surface undulations provide diffusion to minimize focusing of reflected sound.

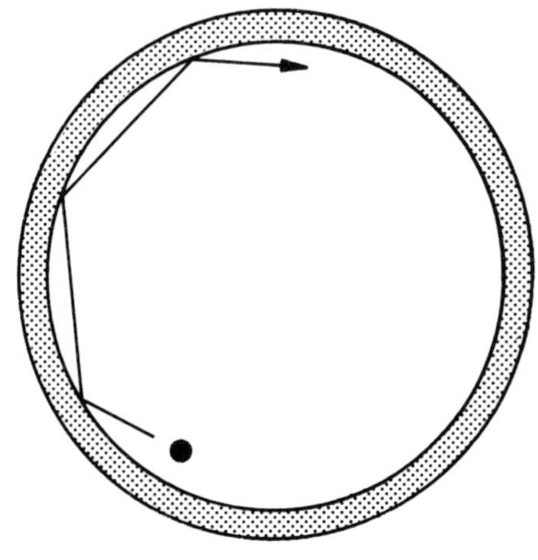

Creep

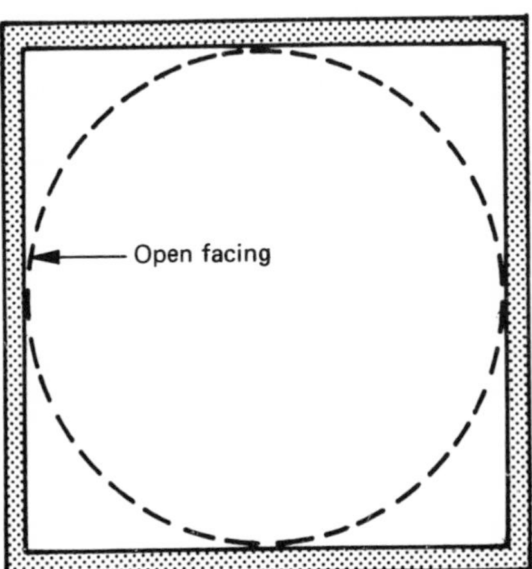

B. Sound-absorbing treatment

Acoustically transparent material (e.g., wood or metal grille) conceals actual enclosure, which can be treated with sound-absorbing material.

NOTE: **A phenomenon called "creep" (i.e., useless sound reflections along smooth concave surfaces) may be present in rooms with curved surfaces and can be controlled by treatments shown above as well.**

ROOM ACOUSTICS: Suspended Sound Reflectors

To reflect sound effectively, a surface should be finished with a hard material such as plaster or plywood and must be greater in size than the wavelength of the sounds of interest. For example, 500 Hz corresponds to a wavelength of about 2 ft. Consequently, to reflect sound of 500 Hz and higher, a panel about four times the wavelength in dimensions at 8 ft should be used. Sound below 500 Hz, however, will bend around the panel and pass by. The hard, sound-reflecting pulpit canopy shown below will provide useful reinforcement of direct sound as well as correct the potential echo conditions from the high ceiling.

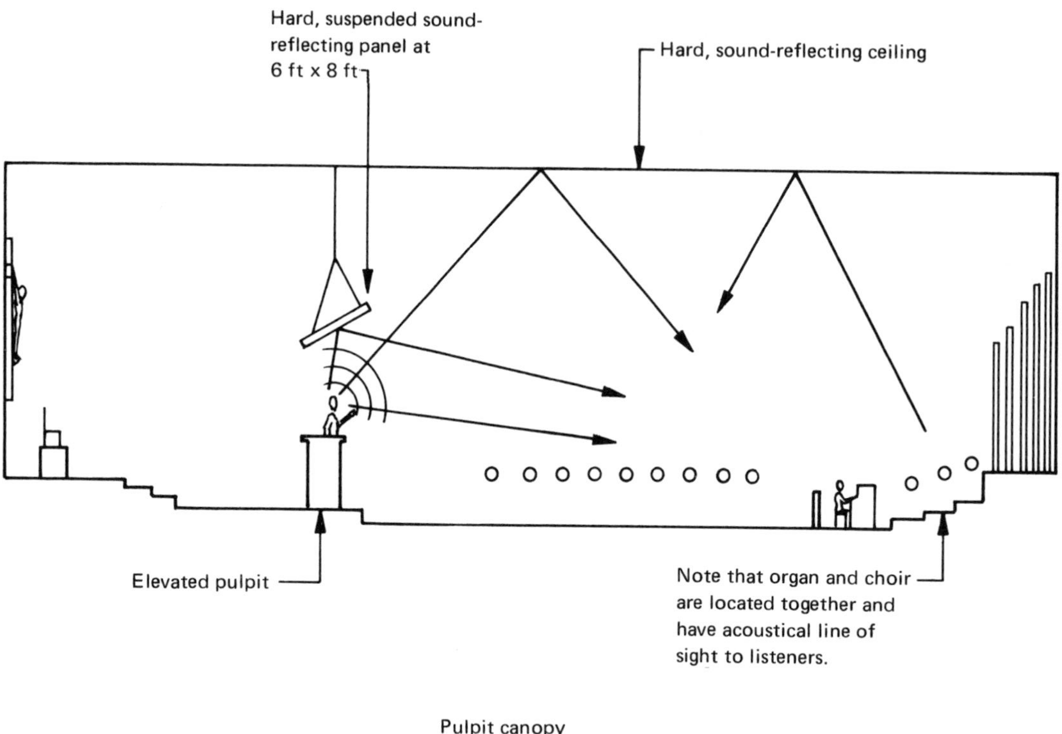

Pulpit canopy

NOTE: **In many churches an interior of wood panels applied to furring is an unintended, effective vibrating panel sound absorber. This can reduce low frequency reverberation considerably, giving an unpleasant "shrill" characteristic to music. If wood is desired, vibrating panel effects can be avoided by adhering the wood to a rigid backup wall.**

ROOM ACOUSTICS: Air Absorption

Air absorption in buildings is not significant at frequencies below 2000 Hz or at very high frequencies above 10,000 Hz. In large spaces the contribution of air absorption should be included in the total room absorption. Note that the curves below show air absorption per 1000 cu ft. It is good practice to use an air absorption coefficient of about 8 at 4000 Hz for large spaces.

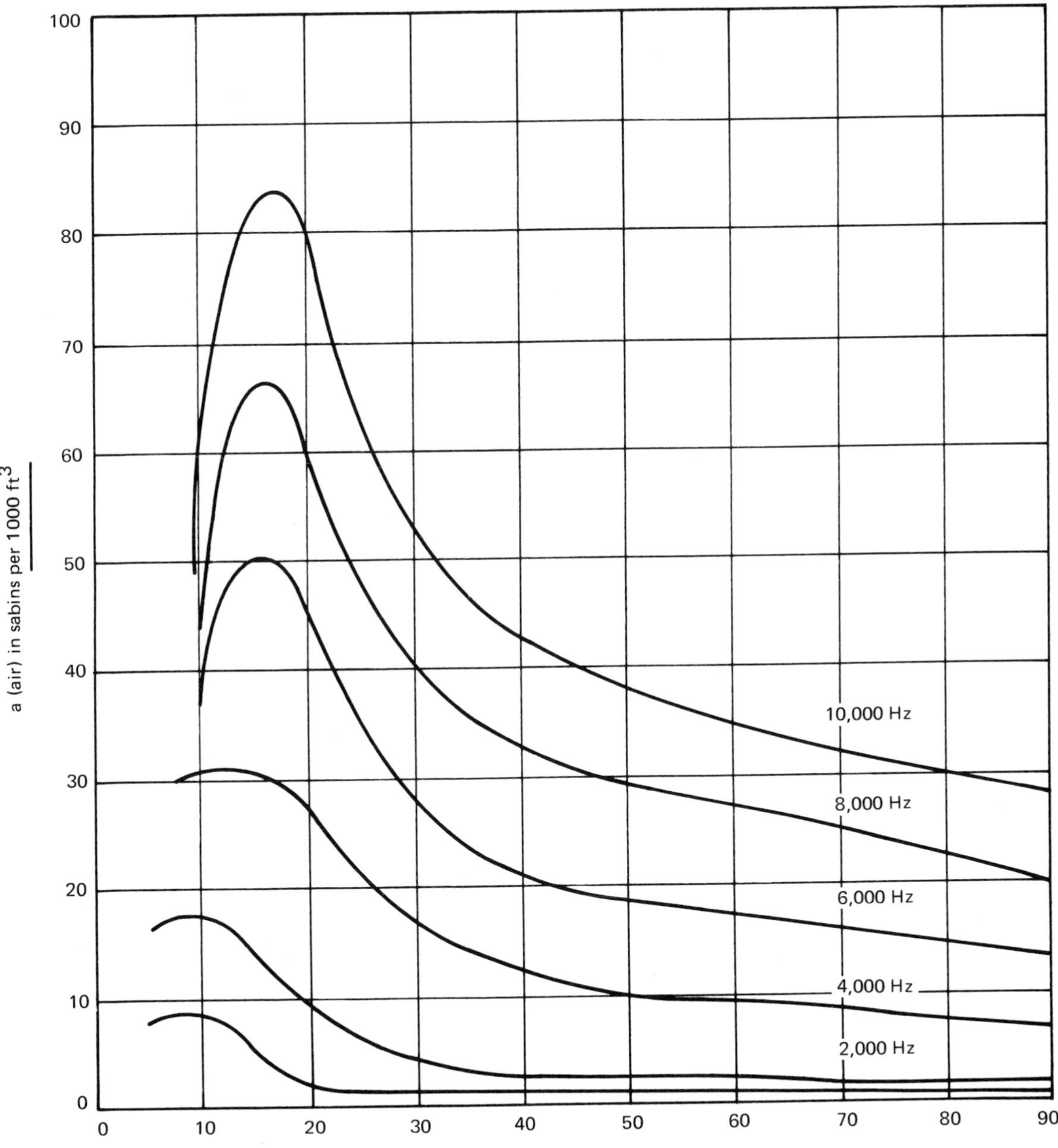

NOTE: Air absorption is the reason thunder "cracks" when heard up close, but "rumbles" when heard at a distance—the high frequencies are attenuated by air!

SUGGESTED AUDITORIUM DESIGN PROCEDURES

Step-by-step design procedures for a typical high school auditorium with seating capacity in the range of 1000 to 2000 are outlined below.

1. **Site Considerations:** Place the auditorium on a quiet exposure, far away from highways, flight paths, etc.
2. **Location within Building:** Use corridors, closets, and quiet "buffer" spaces to isolate the auditorium. No adjacent rehearsal rooms or mechanical equipment rooms under the stage. Avoid occupancy above the auditorium. Select enclosure STCs based on Section 3 procedures. All doors should be solid-core and gasketed around their entire perimeter and should not contain return air grilles. Avoid sliding or roll-up doors. Treat corridors and lobbies with generous amounts of sound-absorbing materials.
3. **Space Use:** Lectures, dramas, orchestra, recitals, etc. Consequently, a full-frequency response sound system will be required.
4. **Volume:** In an auditorium with upholstered seats and a sound-absorbing rear wall for echo control, the average ceiling height H is usually $H = 20T$; where T = mid-frequency reverberation time in seconds. See page 40 for the preferred reverberation time in high school auditoriums. Preliminary shape can be based on architectural considerations; and seating geometry can be arranged to give all the audience good sight lines (e.g., elevated seating layouts) as well as to minimize the distance to the performing area.
5. **Reverberation Time:** Use the Sabine equation $T = 0.05\ V/a$ at 125, 500, and 4000 Hz. This is the formula generally used by the measurement laboratories. Published absorption data on most building materials are intended for use in this formula. The following general guidelines can be used to select finish materials.
   a. **Ceiling:** The central area (about 80%) should be sound-reflecting, e.g., plywood, gypsum board, or plaster on lath; the perimeter along both sides and rear only (about 20%) can be sound-absorbing, e.g., acoustical tile. See sketch on page 41 showing "horseshoe" ceiling treatment pattern.
   b. **Side walls:** Sound-reflecting and diffusing with as many irregularities such as coffers, splays, and undulations as possible.
   c. **Rear wall:** Treat with a "deep" sound-absorbing finish.
   d. **Floor:** Carpet all aisles, except in front of the stage, to aid in footfall noise control. Use fabric-upholstered seats (*never* leatherette) with perforated seat pans. Absorptive seating will help provide stable reverberation. That is, the reverberation time should be nearly the same when the auditorium is full as when it is partially occupied.

   The mid-frequency reverberation time T should be in the range of 1.5 to 1.8 sec. The reverberation time at 125 Hz should be about 1.4T, and at 4000 Hz about 0.8T to give overall fullness and warmth to music. It is important to check the reverberation at high frequencies since too much usually means "harsh" or "rasping" listening conditions. (Don't forget to include air absorption, which may be a significant consideration at high frequencies in large listening spaces.) On the other hand, at low frequencies too much reverberation usually sounds "boomy," whereas too little sounds "shrill."

6. **Ray-diagram Analysis:** Use ray-diagram graphics to properly orient the ceiling and side walls, especially near the proscenium, so they will provide useful sound reflections (i.e., reflections with less than 34 ft difference from the direct sound path). Avoid vaults and domes.
7. **Background Noise Levels:** See Section 3 for the preferred NC levels in auditoriums (i.e., essentially "inaudible"; no masking is wanted here) and adjacent spaces. The mechanical system must meet criteria by controlling air speeds at room registers, and by using internal sound-absorbing duct linings and/or mufflers to prevent duct-borne noise in both the supply and the return air systems.
8. **Stage Enclosure:** Shape for good distribution and diffusion. The stage house reverberation should be approximately equal to that of the auditorium. The orchestra pit, sized at about 20 sq ft per musician, should have a removable sound-absorbing curtain along its rear wall.
9. **Sound-reinforcing System:** Provide a central space just above and in front of the proscenium opening with line of sight to all seats for the sound system's cluster coverage pattern. The height of the proscenium opening (i.e., about 20 to 30 ft at the front of the stage) is generally determined by the need for a stage enclosure. The sound system control console should be located at the rear of the auditorium in a control room which can be opened to the main space by means of an operable plate-glass port. It is important that the system operator be able to hear the sound he is controlling.
10. **Balcony Considerations:** Use balconies to increase seating capacity and/or to reduce the distance to the farthest row of seats. Keep the overhang shallow (i.e., depth less than twice the opening height), slope the soffit, and treat the face with sound-absorbing material.

EXAMPLE PROBLEM – REVERBERATION TIME

Reverberation time computations for a high school auditorium with a volume of 250,000 cu ft are shown below for (a) full and (b) one-half occupancy conditions. An auditorium of this type, which serves many functions, must be designed with a compromise reverberation time. Consequently, select a mid-frequency reverberation time of 1.8 sec, from the chart on page 40. This will be appropriate for the music activities, where blending is needed, and will not be too long for speech activities, especially since a sound system will be used.

	125 Hz	500 Hz	4000 Hz
Preferred reverberation time, sec	2.52	1.80	1.44
Required total absorption, sabins (from Sabine equation)	4960	6944	8680

Material	Area, ft^2	125 Hz Coef.	125 Hz sabins	500 Hz Coef.	500 Hz sabins	4000 Hz Coef.	4000 Hz sabins
a. **Fully Occupied:**							
Ceiling:							
Plaster on lath	8150	0.14	1140	0.06	488	0.03	244
Side walls:							
Plaster on lath	5700	0.14	798	0.06	342	0.03	171
Rear wall:							
2-in. fuzz with open facing	1500	0.60	900	0.82	1230	0.38	570
Aisles:							
Carpet on foam rubber	1600	0.08	128	0.57	912	0.73	1170
Wood	250	0.15	38	0.10	25	0.07	18
Orchestra pit and apron:							
Wood	950	0.15	143	0.10	95	0.07	67
Proscenium opening:							
(Moderately furnished stage)	1100	0.40	440	0.50	550	0.60	660
Air:							
(Coefficient per 1000 cu ft)	....	....	....	....	....	8	2000
Audience:							
Seated in upholstered seats	4500	0.39	1755	0.80	3600	0.87	3920
Total absorption, sabins			5342		7242		8820
b. **One-half Occupied:**							
(Total absorption in auditorium less audience absorption from above)			3587		3642		4900
Fabric well-upholstered seats	2000	0.19	380	0.56	1120	0.59	1180
Audience (includes edge effect)	2500	0.39	975	0.80	2000	0.87	2175
Total absorption, sabins			4942		6762		8255

Since the anticipated normal use condition will be between one-half and full occupancy, the above computations show that the auditorium satisfactorily meets the reverberation time criteria.

ROOM ACOUSTICS: Variable Room Absorption

Adjustable treatments can be used to change the amount of absorption in a room. They enable the reverberation time characteristics to be modified to satisfy the requirements for various room activities.

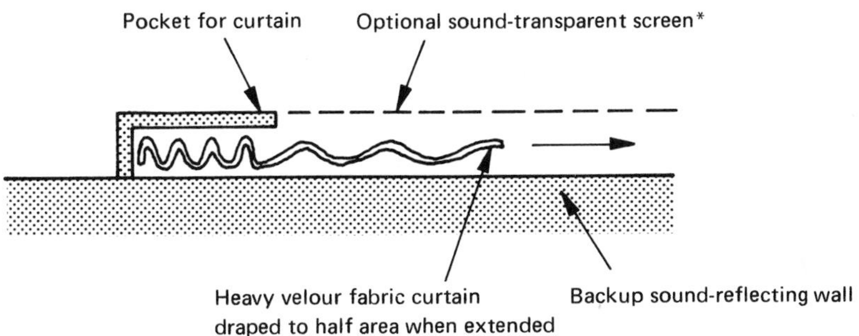

A. Retractable sound-absorbing curtains

***Visually opaque sound-transparent screen is often important since occupants will adjust exposed curtains for visual, not acoustical, reasons.**

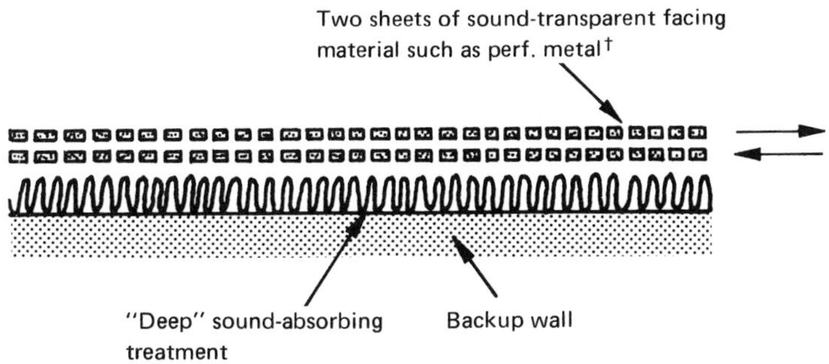

B. Movable facings

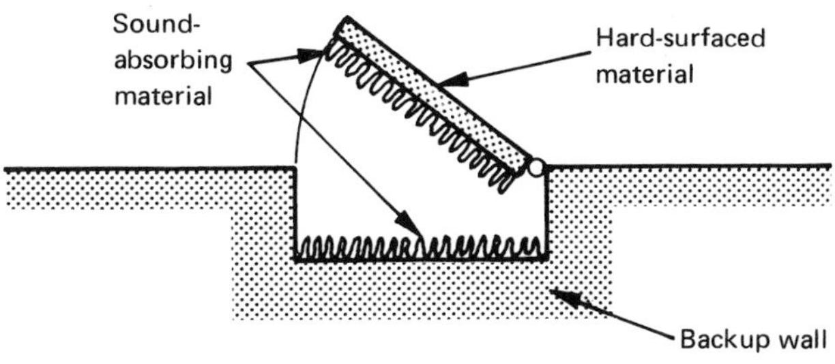

C. Hinged sound-absorbing panels

†Holes line up for condition of maximum absorption.

ROOM ACOUSTICS: Stage Enclosures

On stages where music is important, sound-reflecting and diffusing surfaces are useful near the source of sound. These surfaces help to distribute sound uniformly throughout the listening area and provide good listening conditions on stage, where it is essential that musicians clearly hear each other to perform as a coordinated group. The enclosure can often be designed to be easily erected and dismantled and to be stored compactly without interfering with other stage functional requirements.

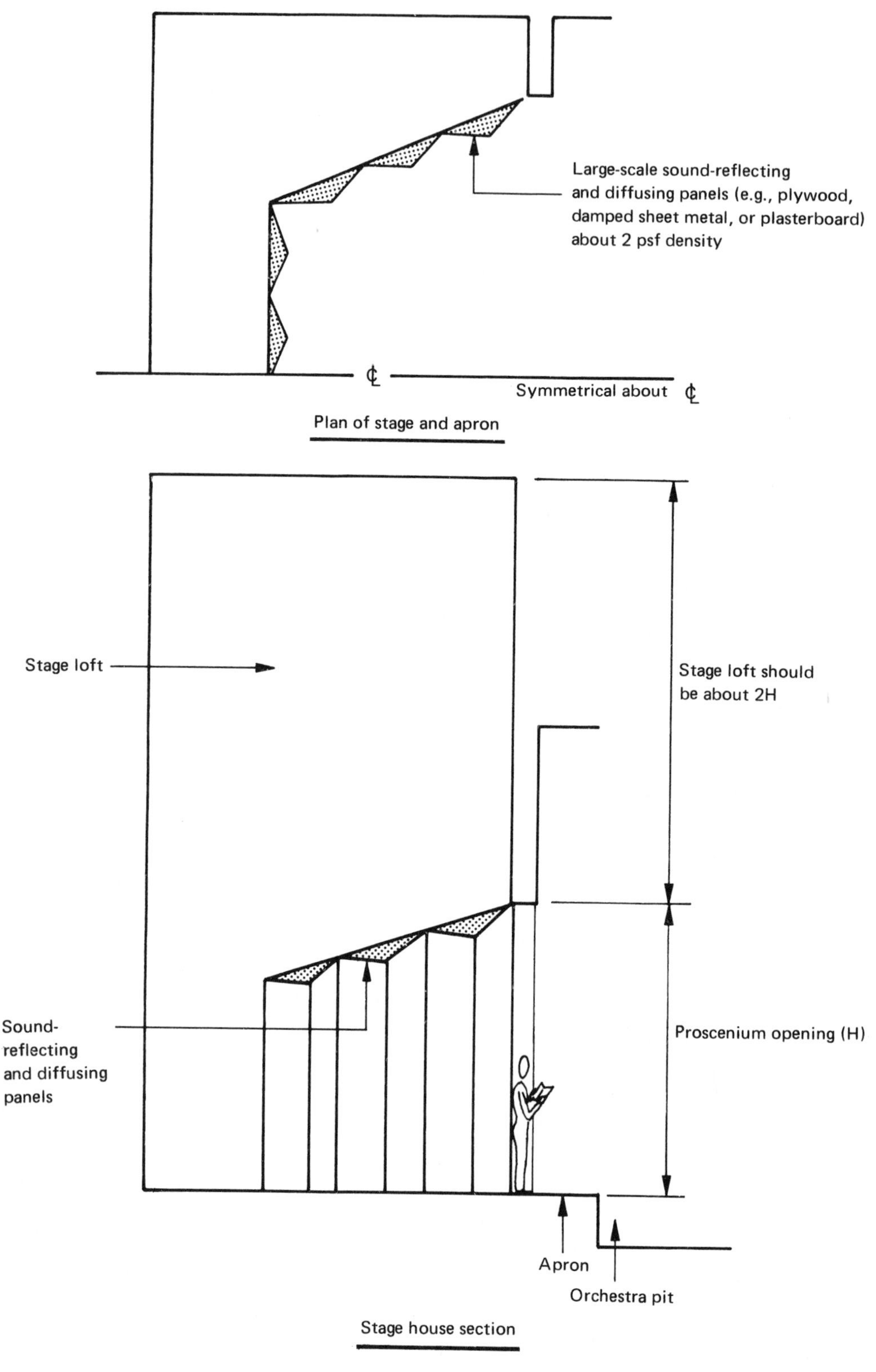

NOTE: Stage house reverberation time should be approx. equal to that of auditorium.

ROOM ACOUSTICS: Balcony Design

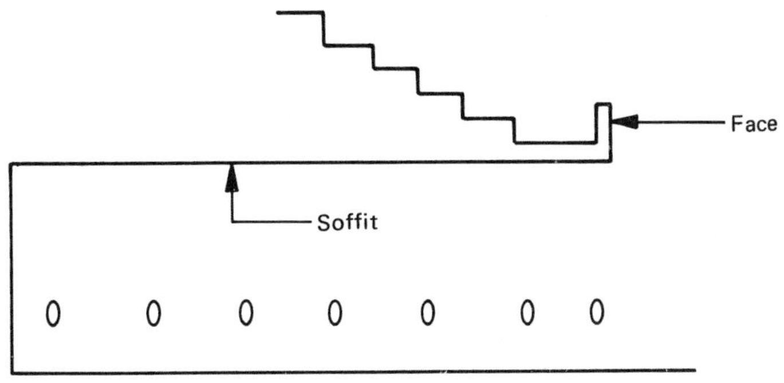

Not this

Persons seated deep under a balcony, or side aisle in a church, cannot receive useful reflected sound from the ceiling and will hear poorly.

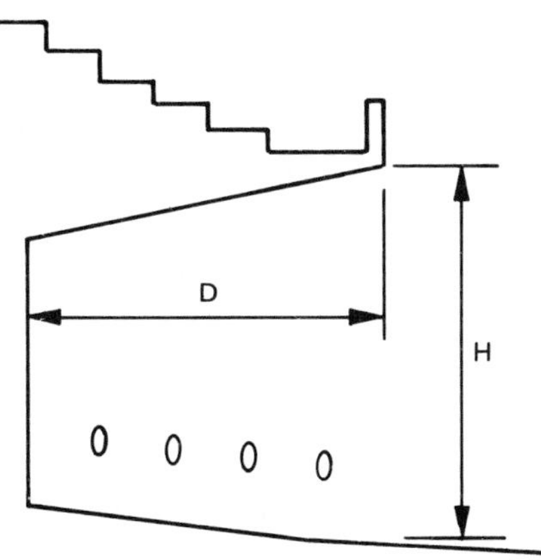

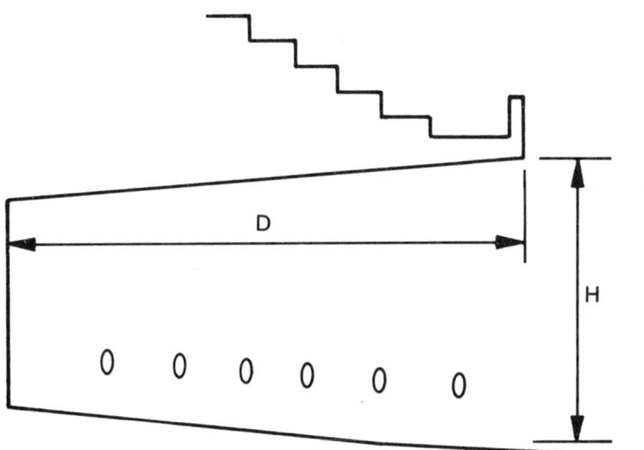

But this

Concert hall

In a concert hall D should not exceed H because symphonic music requires considerable reverberant sound as well as diffusion of the sound.

Opera house

In an opera house D should not exceed 2H.

Motion-picture theater

In motion-picture theaters and similiar facilities D should not exceed 3H; although as in opera houses, 2H is the preferable limit.

NOTE: **Both opera and concert hall balcony soffits, as shown above, should be sloped to direct sound to the heads of the listeners seated underneath. Also, in halls where there is a central loudspeaker system, check to assure that people in the last row have line of sight to the cluster.**

ARTICULATION INDEX

The articulation index (AI) is an objective measure of speech intelligibility which can be calculated from the scores of a group of listeners writing words (or nonsense syllables) read to them from selected lists. The following table relates AI to the percentage of words that a listener hears correctly. For example, if a speaker calls out 100 words, and a listener hears 87 correctly, the words understood would be 87% and the AI above 0.6.

Listening Conditions	Words Understood, %	AI
Excellent	Greater than 84	Above 0.6
Good	62–84	0.4–0.6
Fair	42–61	0.3–0.39
Poor	Below 42	Below 0.3

Two sample lists which can be used to establish an AI for various locations in a room are given below.

Sample List 1

dike	toe
ford	slip
rise	no
creed	heap
strife	folk
pan	pants
bask	grove
fuss	pile
clove	use
smile	hunt
is	death
ride	box
deed	

Sample List 2

fern	mange
then	hit
came	feast
rag	cleanse
pest	such
bar	nook
there (their)	fraud
wheat	rub
hive	bed
not	are
ditch	rat
crash	plush
end	

Note that for speech privacy (cf. Section 4) a low AI is desired. Satisfactory speech privacy means an AI of 0.05 to about 0.15, or words understood from 4 to 14%. This is the converse of communication requirements, where high AIs are essential for satisfactory listening conditions, as shown by the table above.

Reference

L. L. Beranek: *Acoustic Measurements*, Wiley, New York, 1949.

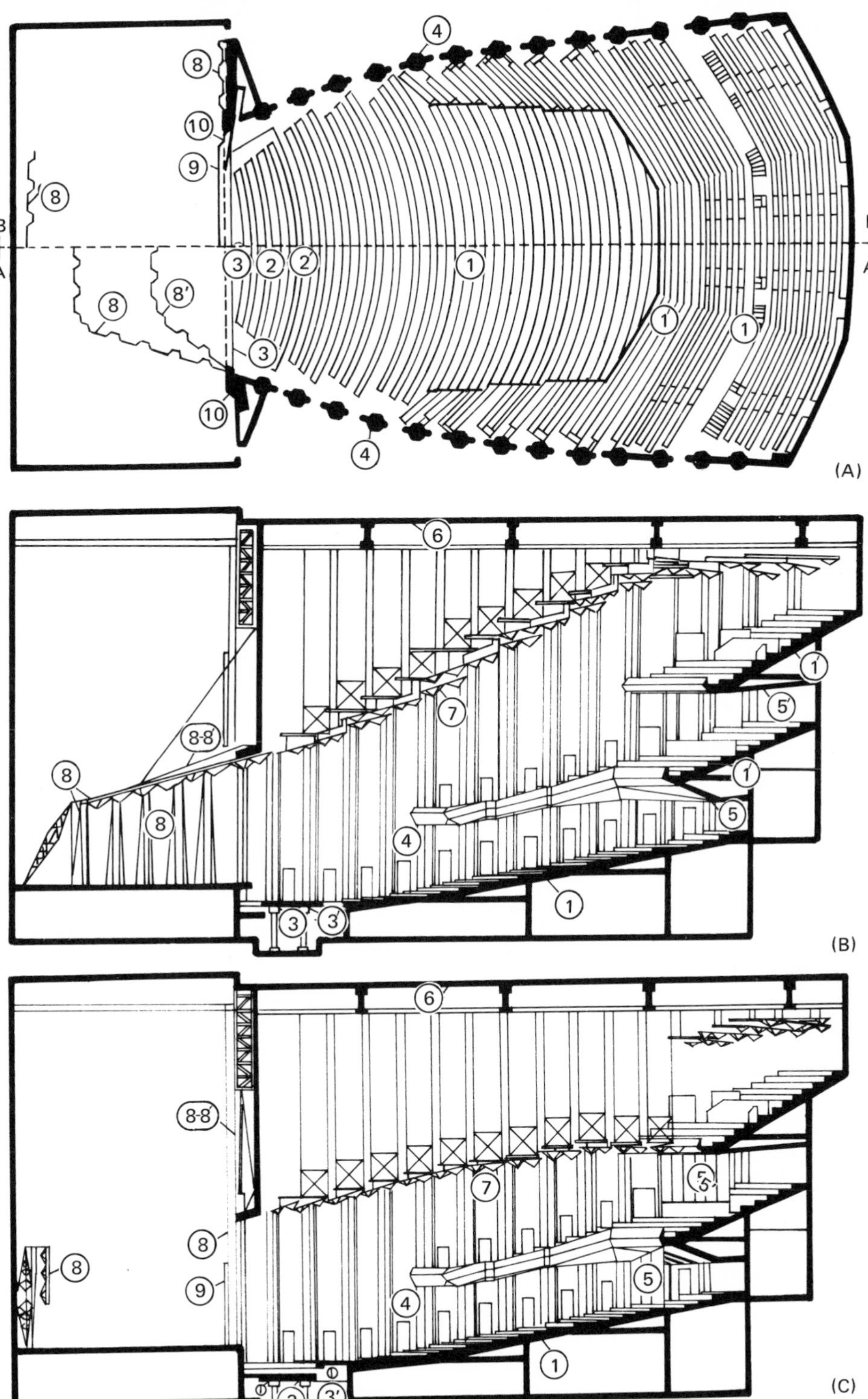

A 3000-seat Concert Hall-Opera House and 1700-seat Recital Hall-theater Combined in One Hall.* (A) – Plan – left stage enclosure in position for 3000-seat concert hall; right, 1700-seat theater, stage enclosure stored. (B) – Longitudinal section, stage enclosure in position for 3000-seat concert hall. (C) – Longitudinal section, stage enclosure stored for 1700-seat theater.

1 – Fixed orchestra seats. 1′ – Fixed balcony seats. 2 and 2′ – Movable orchestra seats. 3 and 3′ – Stage apron elevator. 4 – Fixed auditorium walls. 5 and 5′ – Sound-transparent movable auditorium walls. 6 – Fixed auditorium ceiling. 7 – Movable auditorium ceiling. 8 – Movable stage enclosure, walls, and ceiling. 8′ – Movable stage enclosure walls, recital position. 9 – Stage teaser and tormenter. 10 – Proscenium pullout panels.

Shaded areas are spaces closed off by movable ceiling and rear walls.

*Jones Hall, Houston, Texas
Caudill, Rowlett & Scott, Architects.

Courtesy of dB Magazine.

CHECKLIST FOR EFFECTIVE ROOM ACOUSTICS DESIGN

- The level of the background noise must be sufficiently low to avoid interfering with the intended activities.

- Sound energy must be evenly distributed throughout the space.

- Avoid echoes and focusing effects; and in small rooms having relatively little sound absorption, avoid any shapes which might emphasize certain frequencies (e.g., the ratio of any two of the length, width, and height dimensions should not be a whole number).

- The desired sounds must be sufficiently loud. Shape room surfaces to provide useful sound reflections. If size and space use require a sound-reinforcing system, carefully integrate the system with the room acoustics design.

- Provide the proper reverberation time characteristics. The reverberation time must be long enough to give proper blending of sounds, and yet short enough that there is proper separation of successive sounds necessary for concise audibility. In rooms for both speech and music there is a natural conflict. For example, a long reverberation time is desirable for music so that successive notes blend together, giving richness. However, for speech the reverberation time should be short so that the persistence of one syllable does not blur or mask subsequent syllables.

- Provide a short enough initial-time-delay gap for early reflections. Initial-time-delay gaps should be less than about 30 milliseconds (i.e., a sound path difference of less than 34 ft) to provide useful reinforcement of the direct sound.

SECTION 7

Sound-Reinforcing Systems

SOUND-REINFORCING SYSTEMS: Basic Elements

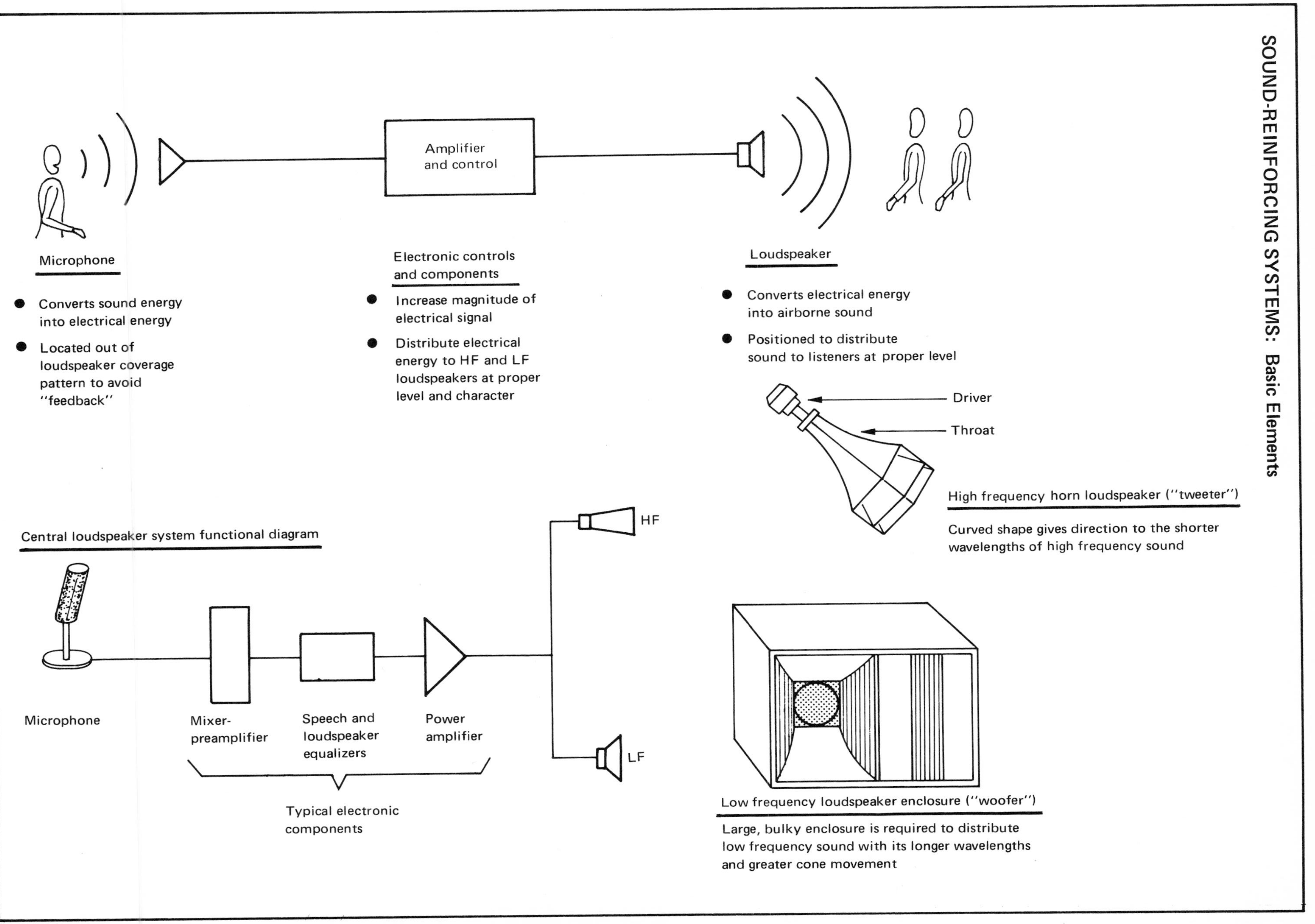

Amplifier
and control
Microphone
Converts sound energy into electrical energy
Located out of loudspeaker coverage pattern to avoid "feedback"
Electronic controls and components
Increase magnitude of electrical signal
Distribute electrical energy to HF and LF loudspeakers at proper level and character
Loudspeaker
Converts electrical energy into airborne sound
Positioned to distribute sound to listeners at proper level
Driver
Throat
High frequency horn loudspeaker ("tweeter")
Curved shape gives direction to the shorter wavelengths of high frequency sound
Central loudspeaker system functional diagram
HF
LF
Microphone
Mixer-preamplifier
Speech and loudspeaker equalizers
Power amplifier
Typical electronic components
Low frequency loudspeaker enclosure ("woofer")
Large, bulky enclosure is required to distribute low frequency sound with its longer wavelengths and greater cone movement

SOUND-REINFORCING SYSTEMS: Central Loudspeaker System

The central loudspeaker system locates the loudspeaker, or cluster of loudspeakers, above the actual source of sound. This system provides maximum realism, as the listener will hear amplified sound from the direction of the natural sound.

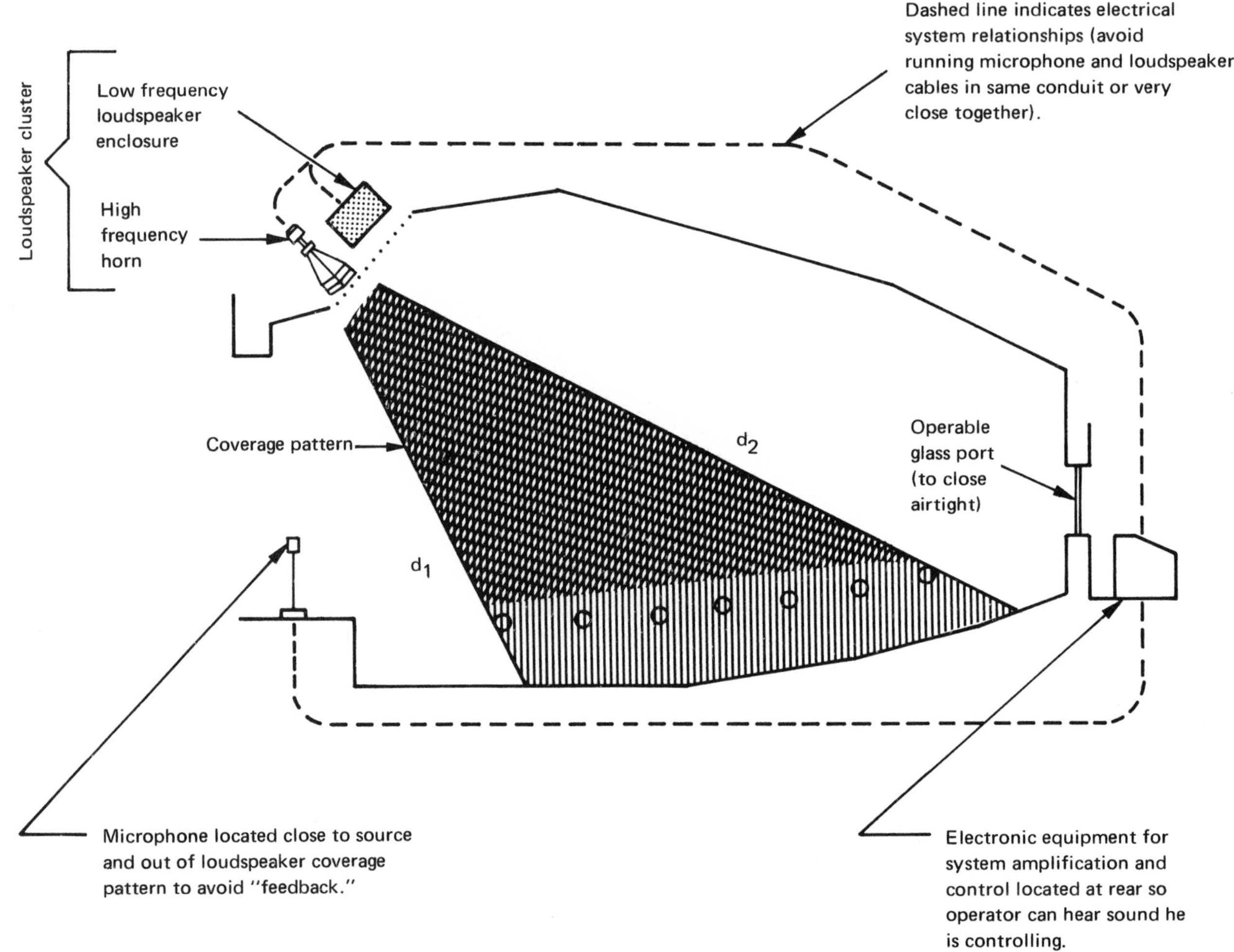

- **Audience should have "line of sight," as high frequency sound is very directional.**
- **Ratio of d_2 to d_1 should not exceed 2 for a high frequency horn (that is, $d_2/d_1 \leqslant 2$).**
- **Sound-reinforcing system should not depend on room surface reflections.**

SOUND REINFORCING SYSTEMS: Distributed Loudspeaker System

The distributed system consists of a number of loudspeakers located overhead. Each supplys low level amplified sound to a small area. It is used in situations where ① the ceiling height is inadequate for a central system, and ② all listeners cannot have line-of-sight to a central cluster (e.g., under a balcony).

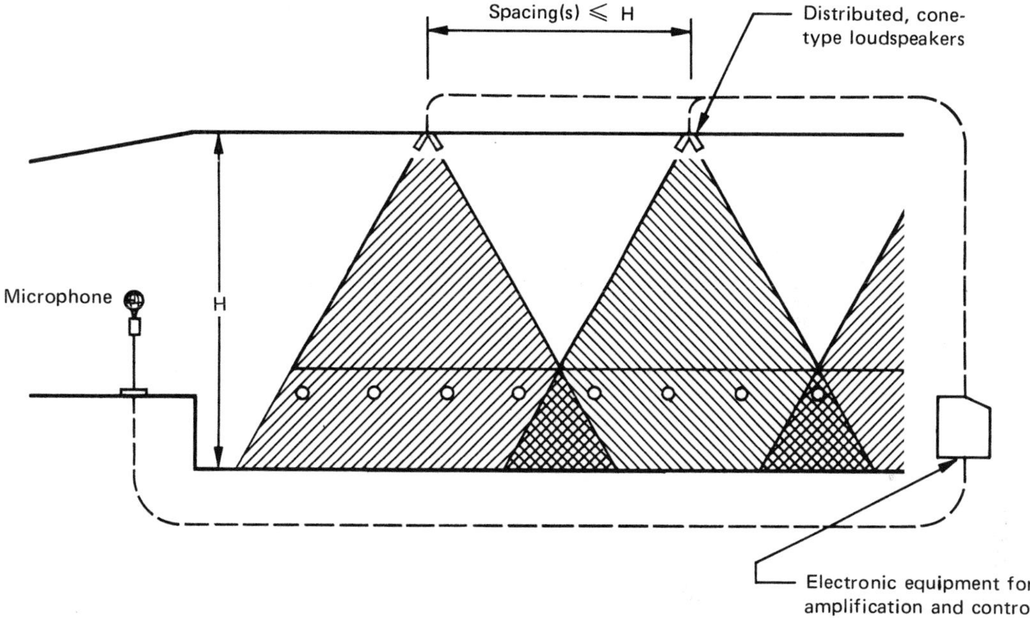

- Ceiling height H should not exceed 20 feet which is about the minimum H for the preferred central system.
- Each listener should receive sound from one loudspeaker.
- Sometimes time delay is required to avoid artificial echo. This can occur when the path difference ("loudspeaker-to-listener" compared to "natural source-to-listener") is greater than about 50 ft.

TIME DELAY

Time delay devices are sometimes required as part of distributed loudspeaker systems in long rooms having low ceiling heights. They can also be used as part of central loudspeaker systems, for example, under a deep balcony overhang. These devices are used to avoid the "artificial echo" or muddying effect on the amplified sound heard almost instantaneously from the loudspeakers located overhead when the natural sound, or sound from a central system cluster, arrives from the front of the room later. The effect is due to the fact that sound in air travels at 1130 fps; whereas the electrical signal to the loudspeakers travels at the speed of light. Therefore, time delay devices are used to introduce a signal delay in the electrical system feeding the distant loudspeakers so that their sound will arrive about the same time as the natural sound.

SYSTEM EQUALIZATION

Sound systems may excite undesirable room resonances, causing "squealing" feedback and uneven distribution of sound, particularly in highly reverberant spaces. The specific resonant frequencies of a space are dependent on its physical dimensions and the sound-absorbing treatment present. However, by using narrowband adjustable electronic filters, it is possible to match a sound system with the room's natural acoustics. For example, those frequencies which excite room resonances can be attenuated or boosted by the correct selection of filters. This procedure is highly specialized and must be performed only by qualified sound system engineers after the system has been installed in the finished space.

SOUND REINFORCING SYSTEMS: Poor Loudspeaker Placements

Poor loudspeaker placements can mean ineffective sound reinforcement. The listening conditions of the placement scheme shown below can actually be improved by shutting the system off! Both arrangements can produce unpleasant effects since they create areas of interference where their coverage patterns overlap.

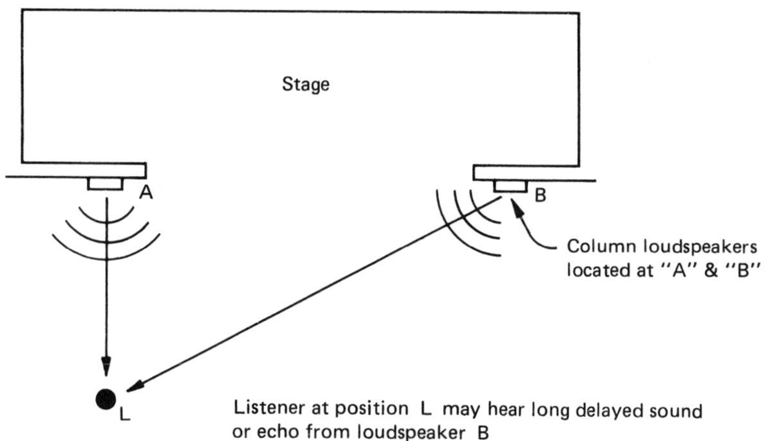

Loudspeakers located on both sides of wide proscenium opening

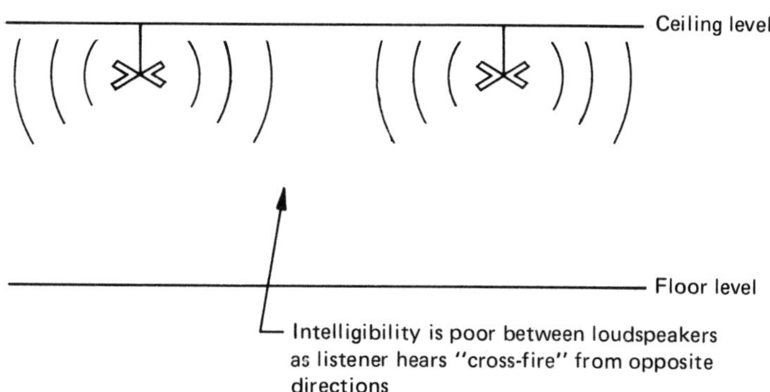

Loudspeakers distributed along circulation space

SOUND-REINFORCING SYSTEMS: Loudspeaker Cluster

A loudspeaker system cluster takes up a great deal of space, and the area in front must be completely sound-transparent, as shown below.

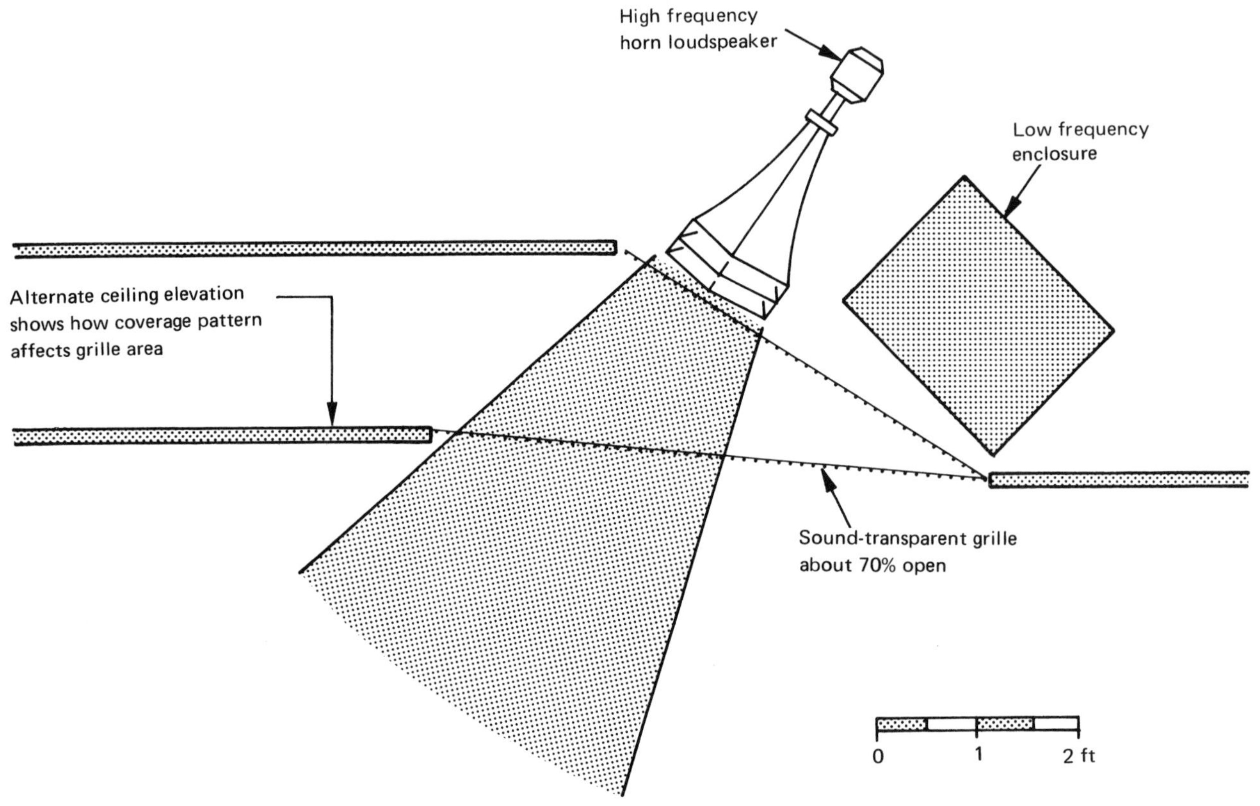

A central loudspeaker cluster

- **Extent of grille area is determined by coverage patterns of the loudspeakers located behind.**
- **Suitable grille materials include monk's cloth, loudspeaker grille cloth, etc.**
- **Solid framing members must be small (less than 5/8" in width) so they will not occlude loudspeaker coverage pattern.**

SOUND-REINFORCING SYSTEMS: Electronic Background Noise System

Electronic masking should be considered only where absolutely necessary—especially in new construction. Distributed loudspeaker systems used for masking applications should be spaced as recommended below.

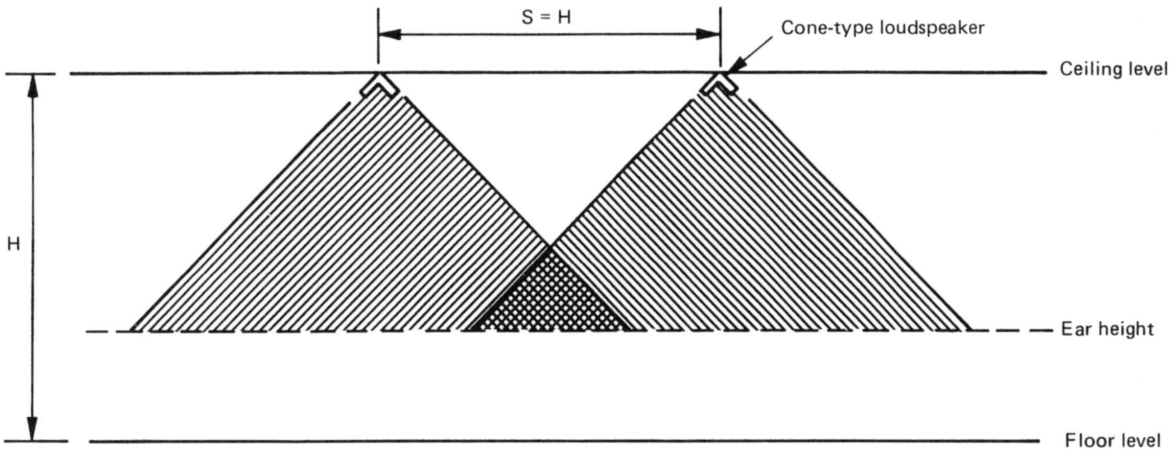

NOTE: Loudspeaker spacing S ideally should about equal room height H as shown here. However, $S = \sqrt{2}\,H$ is the practical spacing limit for uniform coverage.

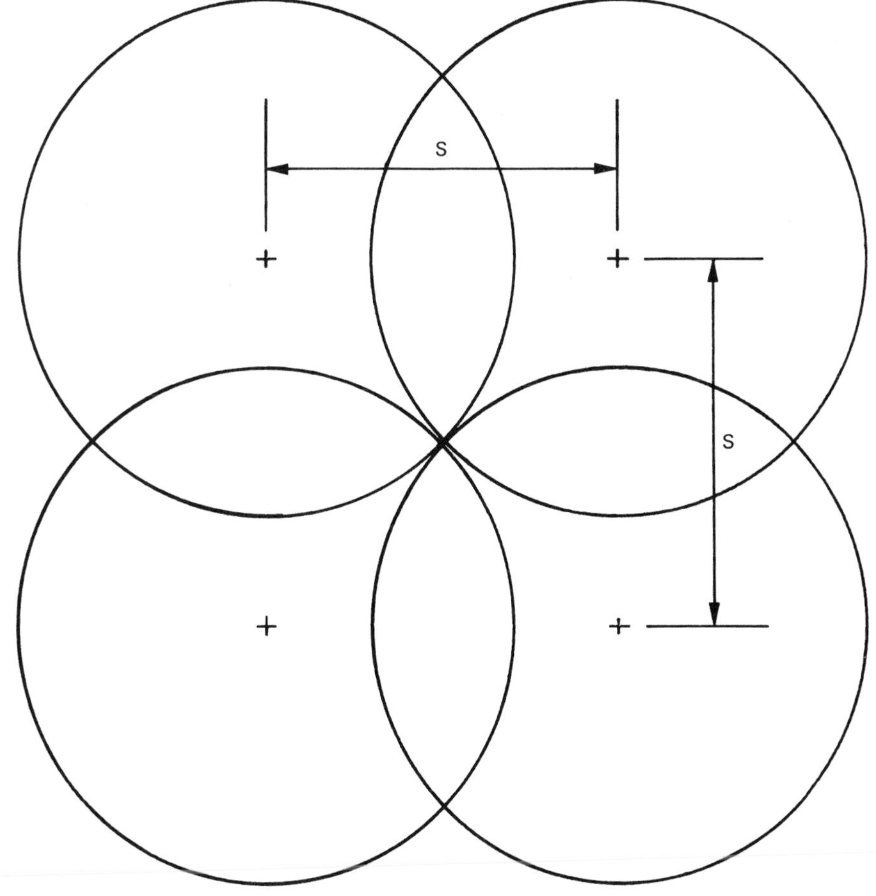

Loudspeaker spacing pattern

CHECKLIST FOR EFFECTIVE USE OF SOUND-REINFORCING SYSTEMS

- A well designed sound-reinforcing system should augment the natural transmission of sound from source to listener. It should be properly integrated with the room acoustics design to provide adequate loudness and good distribution of sound. It should never be used in lieu of good room acoustics design, because it will rarely overcome or correct serious deficiencies; rather, it likely will amplify and exaggerate deficiencies.

- Spaces seating less than 500 will seldom need a sound-reinforcing system; spaces seating 500 to 1000 may need a sound system, depending on the space use; and spaces seating more than 1000 will normally need a sound system, although it may not be used all the time.

- The preferred type of sound-reinforcing system *always* is the central system, in which a loudspeaker, or cluster of loudspeakers, is located directly above the source of sound to give maximum realism as well as intelligibility.

- The other principal type of sound-reinforcing system is the distributed system, in which a large number of loudspeakers, each supplying low-level amplified signals to a small area, are located overhead. The distributed system should be used only when the ceiling height is inadequate to use a central system or when not all listeners can have line of sight to a central loudspeaker.

- Avoid feedback of sound energy from loudspeaker to microphone by careful location of microphones out of the coverage pattern of the loudspeakers. Feedback is the regeneration of a signal between loudspeaker and microphone which is heard as "howling" or "screeching."

- A sound-reinforcing system used only for speech need not reproduce sound down to 63 Hz, so avoid the "bass costs only a little more" sales presentation.

Selected References

1. L. L. Beranek: *Music, Acoustics and Architecture*, Wiley, New York, 1963.
2. L. L. Beranek: *Noise Reduction*, McGraw-Hill, New York, 1960.
3. R. D. Berendt; G. E. Winzer; and C. B. Burroughs: "A Guide to Airborne, Impact, and Structure Borne Noise-Control in Multifamily Dwellings," U.S. Department of Housing and Urban Development, September, 1967.
4. B. F. Day; R. D. Ford; and P. Lord: *Building Acoustics*, American Elsevier, New York, 1969.
5. L. L. Doelle: *Environmental Acoustics*, McGraw-Hill, New York, 1972.
6. W. Furrer: *Room and Building Acoustics and Noise Abatement*, Butterworths, Washington, D.C., 1964.
7. V. O. Knudsen and C. M. Harris: *Acoustical Designing in Architecture*, Wiley, New York, 1950.
8. J. E. Moore: *Design for Noise Reduction*, Architectural Press, London, 1966.
9. R. B. Newman and W. J. Cavanaugh: Acoustics in J. H. Callender (ed.), *Time-saver Standards*, McGraw-Hill, New York, 1966.
10. P. H. Parkins and H. R. Humphreys: *Acoustics, Noise and Buildings*, Faber, London, 1958.
11. H. J. Purkis: *Building Physics: Acoustics*, Pergamon, New York, 1966.
12. W. C. Sabine: *Collected Papers on Acoustics,* Dover, New York, 1964.
13. *Sound Control in Design*, U.S. Gypsum Co., Chicago, 1959.
14. L. F. Yerges: *Sound, Noise, and Vibration Control*, Van Nostrand Reinhold, New York, 1969.

Useful Formulas

GENERAL

Character

1) $f = \frac{1}{T}$

where f = frequency, Hertz (Hz) or cycles/sec (cps)

T = period of one complete vibration, sec

2) $f = \frac{c}{\lambda}$

where f = frequency, Hz

c = speed of sound in air, fps (assume 1130 fps unless otherwise specified)

λ = wavelength, ft

Magnitude

3) $IL = 10 \log \frac{I}{I_0}$

where IL = sound intensity level, decibels (dB)

I = sound intensity, watt/cm^2

I_0 = reference sound intensity, watt/cm^2 (usually taken as 10^{-16} watt/cm^2 or the equivalent 10^{-12} watt/m^2)

4) $SPL = 20 \log \frac{P}{P_0}$

where SPL = sound pressure level, dB

P = sound pressure, dyne/cm^2

P_0 = reference sound pressure, dyne/cm^2 (always taken as 0.0002 dyne/cm^2 or 0.0002 microbar)

NOTE: SPL may be considered equal to IL in most architectural acoustics problems.

Noise Reduction

5) $NR = IL_1 - IL_2$

where NR = noise reduction, dB (or the difference in sound intensity levels between any two conditions)

IL_1 = intensity level, dB, under one condition (usually taken as the higher value)

IL_2 = intensity level, dB, under another condition

NOTE: NR also equals $SPL_1 - SPL_2$ in decibels [see Equation (4)].

6) $NR = 10 \log \frac{I_1}{I_2}$

where NR = [see Equation (5)]

I_1 = sound intensity, watt/cm^2, under one condition

I_2 = sound intensity, watt/cm^2, under another condition

SOUND SOURCE UNDER FREE FIELD CONDITIONS (OUTDOORS)

7) $I = \frac{W}{4\pi d^2}$

where I = sound intensity, watt/cm^2

W = sound power, watts

$\pi = 3.14$

d = distance from sound source, cm

NOTE: If distance is given in English units (i.e., feet), use

$$I = \frac{W}{4\pi d^2 930}$$

8) $PWL = 10 \log \frac{W}{W_0}$

where PWL = sound power level, dB

W = sound power, watts

W_0 = reference sound power, watts (usually taken as 10^{-12} watt)

NOTE: PWL is used by testing laboratories to rate a sound source independently of its environment.

Inverse-square Law:

9) $\frac{I_1}{I_2} = \left(\frac{d_2}{d_1}\right)^2$

where I_1 and I_2 are sound intensities, watt/cm², at distances d_1 and d_2 respectively

10) $NR = 20 \log \frac{d_2}{d_1}$

where NR = [see Equation (5)]

d_2 = distance to one location (usually in feet)

d_1 = distance to another location

NOTE: It can be seen from this expression that the sound level is decreased by 6 dB for each doubling of distance from a point source.

SOUND SOURCE IN REVERBERANT FIELD (INDOORS)

11) $a = \Sigma S\alpha$

or $a = S_1\alpha_1 + S_2\alpha_2 + \cdots + S_n\alpha_n$

where a = total room absorption, sabins, at a given frequency (sum of room surface areas times their respective sound absorption coefficients)

S = surface area, ft²

α = sound absorption coefficient at a given frequency

12) $I = \frac{W}{a}$

where I = sound intensity, watt/cm² (this intensity is the reverberant field intensity and is relatively constant throughout the room in question except very near the source, within about $d = \sqrt{a/4\pi}$, usually 1 to 4 ft, where the sound intensity follows inverse-square law)

W = sound power, watts

a = total room absorption, sabins

NOTE: If absorption is calculated in English units (i.e., square feet) use

$$I = \frac{W}{a930}$$

13) $\frac{I_1}{I_2} = \frac{a_2}{a_1}$

where I_1 and I_2 are reverberant field sound intensities, watt/cm², at two conditions of total room absorption a_1 and a_2 respectively

14) $NR = 10 \log \frac{a_2}{a_1}$

where NR = noise reduction, dB (in this case the reduction in reverberant field intensity level under two different conditions of room absorption, that is, a_2 and a_1)

15) $T = 0.05 \frac{V}{a}$

where T = reverberation time, sec (or time required for sound to diminish 60 dB, or roughly speaking to inaudibility, after source has stopped)

V = room volume, ft^3

a = total room absorption, sabins (if large room, include air absorption)

NOTE: Use the constant 0.16 instead of 0.05 where absorption is calculated in "metric sabins," i.e., surface areas in square meters (m^2).

SOUND TRANSMISSION BETWEEN TWO ROOMS THROUGH A COMMON BARRIER

16) $I_2 = \frac{I_1 \Sigma \tau S}{a_2}$

where I_2 = sound intensity, watt/cm^2, in receiving room (if barrier has only one material, this expression reduces to $I_2 = I_1 \tau S / a_2$)

I_1 = sound intensity, watt/cm^2, in source room

τ = sound transmission coefficient for barrier (τ equals radiated sound power W_2 divided by incident sound power W_1, that is, $\tau = W_2/W_1$)

S = barrier area, ft^2

$\Sigma \tau S$ = sum of areas of each part of barrier times their respective coefficients ($\Sigma \tau S = \tau_1 S_1 + \tau_2 S_2 + \cdots + \tau_n S_n$)

a_2 = total room absorption of receiving room, sabins [see Equation (11)]

17) $NR = 10 \log \frac{a_2}{\Sigma \tau S}$

where NR = noise reduction, dB (in this case the difference in intensity levels, or sound pressure levels, between two rooms)

a_2 = [see Equation (16)]

$\Sigma \tau S$ = [see Equation (16)]

18) $TL = 10 \log \frac{\Sigma S}{\Sigma \tau S}$

where TL = transmission loss of composite barrier, dB [if barrier has only one material, this expression reduces to $TL = 10 \log (1/\tau)$]

ΣS = total area of barrier, ft^2 ($\Sigma S = S_1 + S_2 + \cdots + S_n$)

$\Sigma \tau S$ = [see Equation (16)]

19) $NR = TL + 10 \log \frac{a_2}{S}$

where NR = noise reduction, dB (in this case the difference in intensity levels, or sound pressure levels, between two rooms)

TL = [see Equation (18)]

a_2 = [see Equation (16)]

S = [see Equation (16)]

20) $SPL_1 - NR \leqslant NC$

where SPL_1 = sound pressure level in source room, dB

NR = [see Equation (19)]

NC = background sound pressure level (SPL) of noise criterion, dB (use minimum NC for design purposes)

MECHANICAL SYSTEM NOISE AND VIBRATIONS

21) $f_n = 3.13 \sqrt{\frac{1}{y}}$

where f_n = natural frequency, Hz

y = system static deflection, in.

22) $\frac{\Omega}{f_n} \geqslant 3$

where Ω = forcing (i.e., equipment) frequency, Hz

f_n = [see Equation (21)]

MISCELLANEOUS

23) $NRC = \frac{\alpha_{250} + \alpha_{500} + \alpha_{1000} + \alpha_{2000}}{4}$ with result to nearest 0.05

where NRC = noise reduction coefficient

α = sound absorption coefficients at 250, 500, 1000, and 2000 Hz

24) $f_r = \frac{170}{\sqrt{dG}}$

where f_r = vibrating panel resonant frequency, Hz

d = depth of air space behind panel, in.

G = panel surface density, pounds per square foot (psf)

25) $A = 10 \log \frac{H^2}{R} + 10 \log f - 17$

where A = outdoor barrier attenuation, dB

H = height of barrier above line of sight between source and receiver, ft

R = distance from source to barrier, ft

f = frequency, Hz

26) $H = 20T$

where H = average ceiling height, ft (for auditoriums with upholstered seats and absorptive rear walls)

T = mid-frequency reverberation time, sec

27) $TL = 19 + 20 \log G$

where TL = transmission loss of single homogeneous panel at 400 Hz, dB

G = panel surface density, psf

Use of Logarithm Tables

Familiarization with logarithm tables is needed, not only as a means of avoiding drudgery with arduous multiplications and divisions, but because the log function is basic to most acoustical computations.

The first step in finding the log of a number is to express it as a digit (from 1 to 9) multiplied by 10 to a power n. For example:

	Number		*Digit*		*10^n*
a)	1,723,000	=	1.723	X	10^6
b)	0.0298	=	2.98	X	10^{-2}
c)	1803	=	1.803	X	10^3
d)	0.85	=	8.5	X	10^{-1}

A logarithm, in general, consists of the sum of two parts – a *characteristic* and a *mantissa*. The characteristic is the power to which 10 is raised, and the mantissa if found directly from tables.

The logarithms for the above example numbers can be found as follows:

	N				*Characteristic*		*Mantissa*		*Logarithm*
a)	1.723	X	10^6	=	6	+	0.2362	=	6.2362
b)	2.98	X	10^{-2}	=	−2	+	0.4742	=	−1.5258
c)	1.803	X	10^3	=	3	+	0.2560	=	3.2560
d)	8.5	X	10^{-1}	=	−1	+	0.9294	=	−0.0706

NOTE: Example a) can be expressed as $\log(1.723 \times 10^6) = \log(1.723) + \log(10^6) = \log(1.723) + \log(1 \times 10^6) = 0.2362 + 6.0 = 6.2362$ by using the basic property of logs that $\log xy = \log x + \log y$.

FOUR-PLACE LOGARITHMS TO BASE 10

The logarithms to base 10 of numbers between 1 and 10, correct to four places, are given in the tables shown on this and the following page.

If the decimal point in the number is moved n places to the right (or left), the value of n (or —n) is added to the logarithm, thus:

log 3.14 = 0.4969
log 314. = 0.4969 + 2 or 2.4969
log .0314 = 0.4969 — 2, which may be written $\bar{2}.4969$ or 8.4969 — 10

If the given number has more than four significant figures, it should be reduced to four figures, since those beyond four figures will not affect the result in four-place computations.

The logarithm of a number having four significant figures must be interpolated by adding to the logarithm of the three-figure number, the amount under the fourth figure, as read in the proportional parts section of the table.

Thus, the logarithm of 3.1416 is found as follows:

a. Reduce the number to four significant figures: 3.142
b. The log of 3.14 is .4969
c. The value of the proportional part under 2 (the fourth figure) is 3
d. Then, the log 3.142 = 0.4969 + .0003 or 0.4972

Natural logarithms: Many calculations make use of natural logarithms (Base e = 2.7183). To convert base 10 (common) logarithms to natural logarithms, multiply the value for the former by 2.30258.

Natural logarithms are also called hyperbolic or Napierian logarithms.

$$\log ab = \log a + \log b \qquad \log a^n = n \log a$$

$$\log \frac{a}{b} = \log a - \log b \qquad \log \sqrt[a]{a} = \frac{\log a}{n}$$

N	0	1	2	3	4	5	6	7	8	9	Proportional Parts								
											1	2	3	4	5	6	7	8	9
1.0	0000	0043	0086	0128	0170	0212	0253	0294	0334	0374	4	8	12	17	21	25	29	33	37
1.1	0414	0453	0492	0531	0569	0607	0645	0682	0719	0755	4	8	11	15	19	23	26	30	34
1.2	0792	0828	0864	0899	0934	0969	1004	1038	1072	1106	3	7	10	14	17	21	24	28	31
1.3	1139	1173	1206	1239	1271	1303	1335	1367	1399	1430	3	6	10	13	16	19	23	26	29
1.4	1461	1492	1523	1553	1584	1614	1644	1673	1703	1732	3	6	9	12	15	18	21	24	27
1.5	1761	1790	1818	1847	1875	1903	1931	1959	1987	2014	3	6	8	11	14	17	20	22	25
1.6	2041	2068	2095	2122	2148	2175	2201	2227	2253	2279	3	5	8	11	13	16	18	21	24
1.7	2304	2330	2355	2380	2405	2430	2455	2480	2504	2529	2	5	7	10	12	15	17	20	22
1.8	2553	2577	2601	2625	2648	2672	2695	2718	2742	2765	2	5	7	9	12	14	16	19	21
1.9	2788	2810	2833	2856	2878	2900	2923	2945	2967	2989	2	4	7	9	11	13	16	18	20
2.0	3010	3032	3054	3075	3096	3118	3139	3160	3181	3201	2	4	6	8	11	13	15	17	19
2.1	3222	3243	3263	3284	3304	3324	3345	3365	3385	3404	2	4	6	8	10	12	14	16	18
2.2	3424	3444	3464	3483	3502	3522	3541	3560	3579	3598	2	4	6	8	10	12	14	15	17
2.3	3617	3636	3655	3674	3692	3711	3729	3747	3766	3784	2	4	6	7	9	11	13	15	17
2.4	3802	3820	3838	3856	3874	3892	3909	3927	3945	3962	2	4	5	7	9	11	12	14	16
2.5	3979	3997	4014	4031	4048	4065	4082	4099	4116	4133	2	3	5	7	9	10	12	14	15
2.6	4150	4166	4183	4200	4216	4232	4249	4265	4281	4298	2	3	5	7	8	10	11	13	15
2.7	4314	4330	4346	4362	4378	4393	4409	4425	4440	4456	2	3	5	6	8	9	11	13	14
2.8	4472	4487	4502	4518	4533	4548	4564	4579	4594	4609	2	3	5	6	8	9	11	12	14
2.9	4624	4639	4654	4669	4683	4698	4713	4728	4742	4757	1	3	4	6	7	9	10	12	13
3.0	4771	4786	4800	4814	4829	4843	4857	4871	4886	4900	1	3	4	6	7	9	10	11	13
3.1	4914	4928	4942	4955	4969	4983	4997	5011	5024	5038	1	3	4	6	7	8	10	11	12
3.2	5051	5065	5079	5092	5105	5119	5132	5145	5159	5172	1	3	4	5	7	8	9	11	12
3.3	5185	5198	5211	5224	5237	5250	5263	5276	5289	5302	1	3	4	5	6	8	9	10	12
3.4	5315	5328	5340	5353	5366	5378	5391	5403	5416	5428	1	3	4	5	6	8	9	10	11
3.5	5441	5453	5465	5478	5490	5502	5514	5527	5539	5551	1	2	4	5	6	7	9	10	11
3.6	5563	5575	5587	5599	5611	5623	5635	5647	5658	5670	1	2	4	5	6	7	8	10	11
3.7	5682	5694	5705	5717	5729	5740	5752	5763	5775	5786	1	2	3	5	6	7	8	9	10
3.8	5798	5809	5821	5832	5843	5855	5866	5877	5888	5899	1	2	3	5	6	7	8	9	10
3.9	5911	5922	5933	5944	5955	5966	5977	5988	5999	6010	1	2	3	4	5	7	8	9	10

FOUR-PLACE LOGARITHMS TO BASE 10 – continued

N	0	1	2	3	4	5	6	7	8	9	Proportional Parts 1	2	3	4	5	6	7	8	9
4.0	6021	6031	6042	6053	6064	6075	6085	6096	6107	6117	1	2	3	4	5	6	8	9	10
4.1	6128	6138	6149	6160	6170	6180	6191	6201	6212	6222	1	2	3	4	5	6	7	8	9
4.2	6232	6243	6253	6263	6274	6284	6294	6304	6314	6325	1	2	3	4	5	6	7	8	9
4.3	6335	6345	6355	6365	6375	6385	6395	6405	6415	6425	1	2	3	4	5	6	7	8	9
4.4	6435	6444	6454	6464	6474	6484	6493	6503	6513	6522	1	2	3	4	5	6	7	8	9
4.5	6532	6542	6551	6561	6571	6580	6590	6599	6609	6618	1	2	3	4	5	6	7	8	9
4.6	6628	6637	6646	6656	6665	6675	6684	6693	6702	6712	1	2	3	4	5	6	7	7	8
4.7	6721	6730	6739	6749	6758	6767	6776	6785	6794	6803	1	2	3	4	5	5	6	7	8
4.8	6812	6821	6830	6839	6848	6857	6866	6875	6884	6893	1	2	3	4	4	5	6	7	8
4.9	6902	6911	6920	6928	6937	6946	6955	6964	6972	6981	1	2	3	4	4	5	6	7	8
5.0	6990	6998	7007	7016	7024	7033	7042	7050	7059	7067	1	2	3	3	4	5	6	7	8
5.1	7076	7084	7093	7101	7110	7118	7126	7135	7143	7152	1	2	3	3	4	5	6	7	8
5.2	7160	7168	7177	7185	7193	7202	7210	7218	7226	7235	1	2	2	3	4	5	6	7	7
5.3	7243	7251	7259	7267	7275	7284	7292	7300	7308	7316	1	2	2	3	4	5	6	6	7
5.4	7324	7332	7340	7348	7356	7364	7372	7380	7388	7396	1	2	2	3	4	5	6	6	7
5.5	7404	7412	7419	7427	7435	7443	7451	7459	7466	7474	1	2	2	3	4	5	5	6	7
5.6	7482	7490	7497	7505	7513	7520	7528	7536	7543	7551	1	2	2	3	4	5	5	6	7
5.7	7559	7566	7574	7582	7589	7597	7604	7612	7619	7627	1	2	2	3	4	5	5	6	7
5.8	7634	7642	7649	7657	7664	7672	7679	7686	7694	7701	1	1	2	3	4	4	5	6	7
5.9	7709	7716	7723	7731	7738	7745	7752	7760	7767	7774	1	1	2	3	4	4	5	6	7
6.0	7782	7789	7796	7803	7810	7818	7825	7832	7839	7846	1	1	2	3	4	4	5	6	6
6.1	7853	7860	7868	7875	7882	7889	7896	7903	7910	7917	1	1	2	3	4	4	5	6	6
6.2	7924	7931	7938	7945	7952	7959	7966	7973	7980	7987	1	1	2	3	3	4	5	6	6
6.3	7993	8000	8007	8014	8021	8028	8035	8041	8048	8055	1	1	2	3	3	4	5	5	6
6.4	8062	8069	8075	8082	8089	8096	8102	8109	8116	8122	1	1	2	3	3	4	5	5	6
6.5	8129	8136	8142	8149	8156	8162	8169	8176	8182	8189	1	1	2	3	3	4	5	5	6
6.6	8195	8202	8209	8215	8222	8228	8235	8241	8248	8254	1	1	2	3	3	4	5	5	6
6.7	8261	8267	8274	8280	8287	8293	8299	8306	8312	8319	1	1	2	3	3	4	5	5	6
6.8	8325	8331	8338	8344	8351	8357	8363	8370	8376	8382	1	1	2	3	3	4	4	5	6
6.9	8388	8395	8401	8407	8414	8420	8426	8432	8439	8445	1	1	2	2	3	4	4	5	6
7.0	8451	8457	8463	8470	8476	8482	8488	8494	8500	8506	1	1	2	2	3	4	4	5	6
7.1	8513	8519	8525	8531	8537	8543	8549	8555	8561	8567	1	1	2	2	3	4	4	5	5
7.2	8573	8579	8585	8591	8597	8603	8609	8615	8621	8627	1	1	2	2	3	4	4	5	5
7.3	8633	8639	8645	8651	8657	8663	8669	8675	8681	8686	1	1	2	2	3	4	4	5	5
7.4	8692	8698	8704	8710	8716	8722	8727	8733	8739	8745	1	1	2	2	3	4	4	5	5
7.5	8751	8756	8762	8768	8774	8779	8785	8791	8797	8802	1	1	2	2	3	3	4	5	5
7.6	8808	8814	8820	8825	8831	8837	8842	8848	8854	8859	1	1	2	2	3	3	4	5	5
7.7	8865	8871	8876	8882	8887	8893	8899	8904	8910	8915	1	1	2	2	3	3	4	4	5
7.8	8921	8927	8932	8938	8943	8949	8954	8960	8965	8971	1	1	2	2	3	3	4	4	5
7.9	8976	8982	8987	8993	8998	9004	9009	9015	9020	9025	1	1	2	2	3	3	4	4	5
8.0	9031	9036	9042	9047	9053	9058	9063	9069	9074	9079	1	1	2	2	3	3	4	4	5
8.1	9085	9090	9096	9101	9106	9112	9117	9122	9128	9133	1	1	2	2	3	3	4	4	5
8.2	9138	9143	9149	9154	9159	9165	9170	9175	9180	9186	1	1	2	2	3	3	4	4	5
8.3	9191	9196	9201	9206	9212	9217	9222	9227	9232	9238	1	1	2	2	3	3	4	4	5
8.4	9243	9248	9253	9258	9263	9269	9274	9279	9284	9289	1	1	2	2	3	3	4	4	5
8.5	9294	9299	9304	9309	9315	9320	9325	9330	9335	9340	1	1	2	2	3	3	4	4	5
8.6	9345	9350	9355	9360	9365	9370	9375	9380	9385	9390	1	1	2	2	3	3	4	4	5
8.7	9395	9400	9405	9410	9415	9420	9425	9430	9435	9440	0	1	1	2	2	3	3	4	4
8.8	9445	9450	9455	9460	9465	9469	9474	9479	9484	9489	0	1	1	2	2	3	3	4	4
8.9	9494	9499	9504	9509	9513	9518	9523	9528	9533	9538	0	1	1	2	2	3	3	4	4
9.0	9542	9547	9552	9557	9562	9566	9571	9576	9581	9586	0	1	1	2	2	3	3	4	4
9.1	9590	9595	9600	9605	9609	9614	9619	9624	9628	9633	0	1	1	2	2	3	3	4	4
9.2	9638	9643	9647	9652	9657	9661	9666	9671	9675	9680	0	1	1	2	2	3	3	4	4
9.3	9685	9689	9694	9699	9703	9708	9713	9717	9722	9727	0	1	1	2	2	3	3	4	4
9.4	9731	9736	9741	9745	9750	9754	9759	9763	9768	9773	0	1	1	2	2	3	3	4	4
9.5	9777	9782	9786	9791	9795	9800	9805	9809	9814	9818	0	1	1	2	2	3	3	4	4
9.6	9823	9827	9832	9836	9841	9845	9850	9854	9859	9863	0	1	1	2	2	3	3	4	4
9.7	9868	9872	9877	9881	9886	9890	9894	9899	9903	9908	0	1	1	2	2	3	3	4	4
9.8	9912	9917	9921	9926	9930	9934	9939	9943	9948	9952	0	1	1	2	2	3	3	4	4
9.9	9956	9961	9965	9969	9974	9978	9983	9987	9991	9996	0	1	1	2	2	3	3	3	4
N	0	1	2	3	4	5	6	7	8	9	1	2	3	4	5	6	7	8	9

Courtesy of Crane Company

DECIBEL CALCUALTION CHART

If the value of x or 1/x is known, the value of 10 log x can be found on the center decibel scale of the chart.

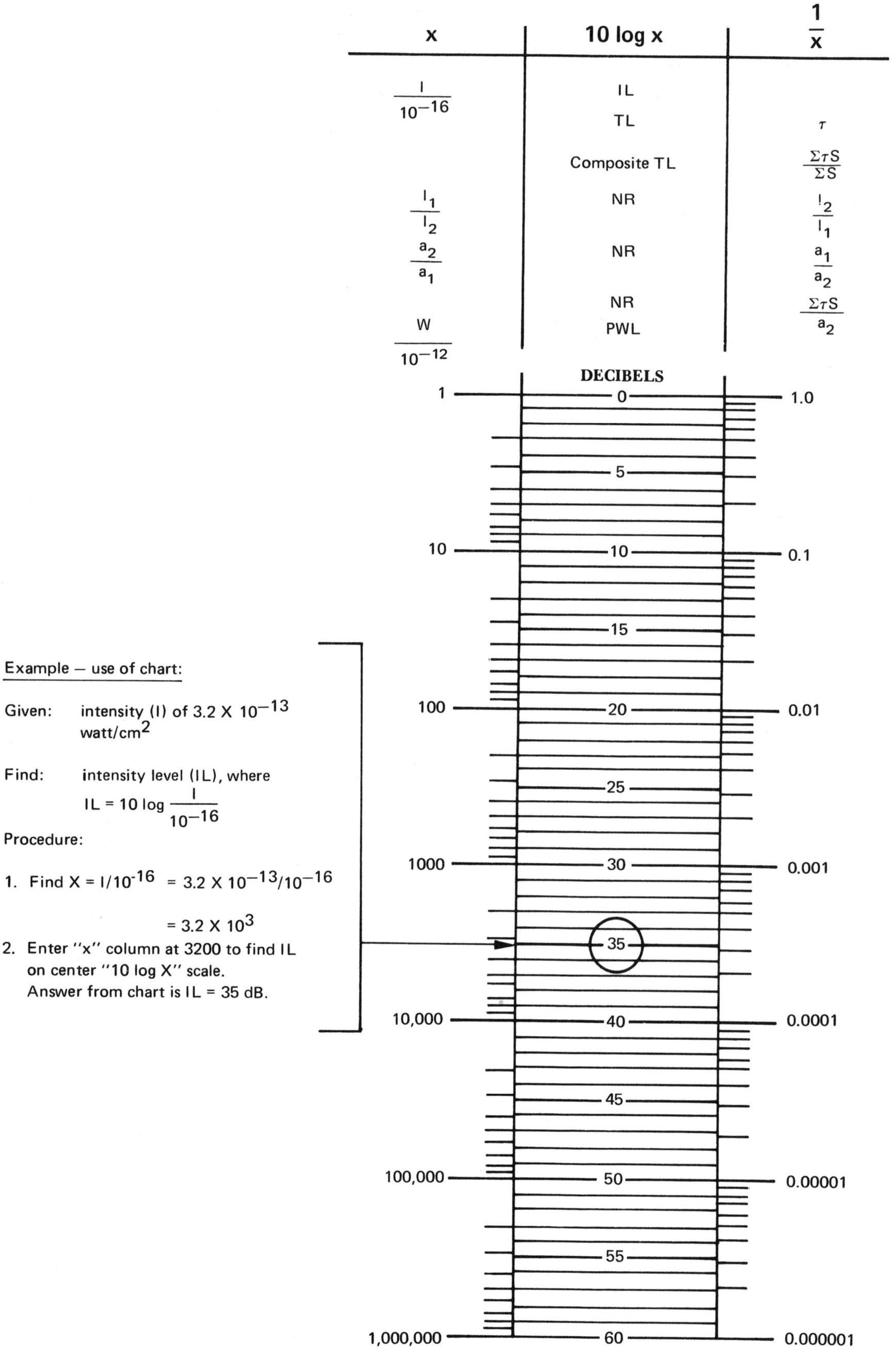

Ref.: R. B. Newman and W. J. Cavanaugh, Acoustics, in J. H. Callender (ed.), Time-saver Standards, 4th ed., McGraw Hill, New York, 1966.

Preferred Noise Criteria

Background noise from airflow at mechanical system room registers or from electronic masking systems should be continuous so that people will hardly notice it. Its frequency spectrum should not only satisfy masking requirements but have a pleasant tonal quality as well. Experience shows that a background noise whose spectrum conforms to the NC curve shape will not sound pleasant as it will contain low frequency "rumble" and high frequency "hissiness." Primarily for this reason a new rating system has been developed to define acceptable background noise. This system consists of standard rating curves called preferred noise criteria (PNC) that have sound pressure levels lower than the NC curves at both the low and high frequencies. PNC curves can be used to evaluate existing noise situations and to specify design goals for acoustical backgrounds in rooms. The PNC curves and rating procedures are given on the following pages. For suggested background PNC ranges appropriate to various indoor functional activities (similar to the recommended NC ranges on page 86) see pages 584–586, L. L. Beranek: *Noise and Vibration Control*, McGraw-Hill, New York, 1971.

Reference

L. L. Beranek; W. E. Blazier; and J. J. Figwer: Preferred Noise Criterion (PNC) Curves and Their Application to Rooms, *J. Acoust. Soc. Amer.*, Vol. 50, no. 5, November 1971.

APPENDIX C: Preferred Noise Criteria (PNC) Curves

In general, PNC curves should only be used to rate noises whose sound spectra have the shape of the PNC curves. The rating of a PNC level for a noise is found by comparing its sound spectrum to the PNC curves shown below. Its SPLs may exceed a PNC curve by up to +2 dB in only one band if in the two adjacent bands the level is not more than −1 dB below the PNC curve. The PNC rating then is the highest PNC curve meeting these requirements.

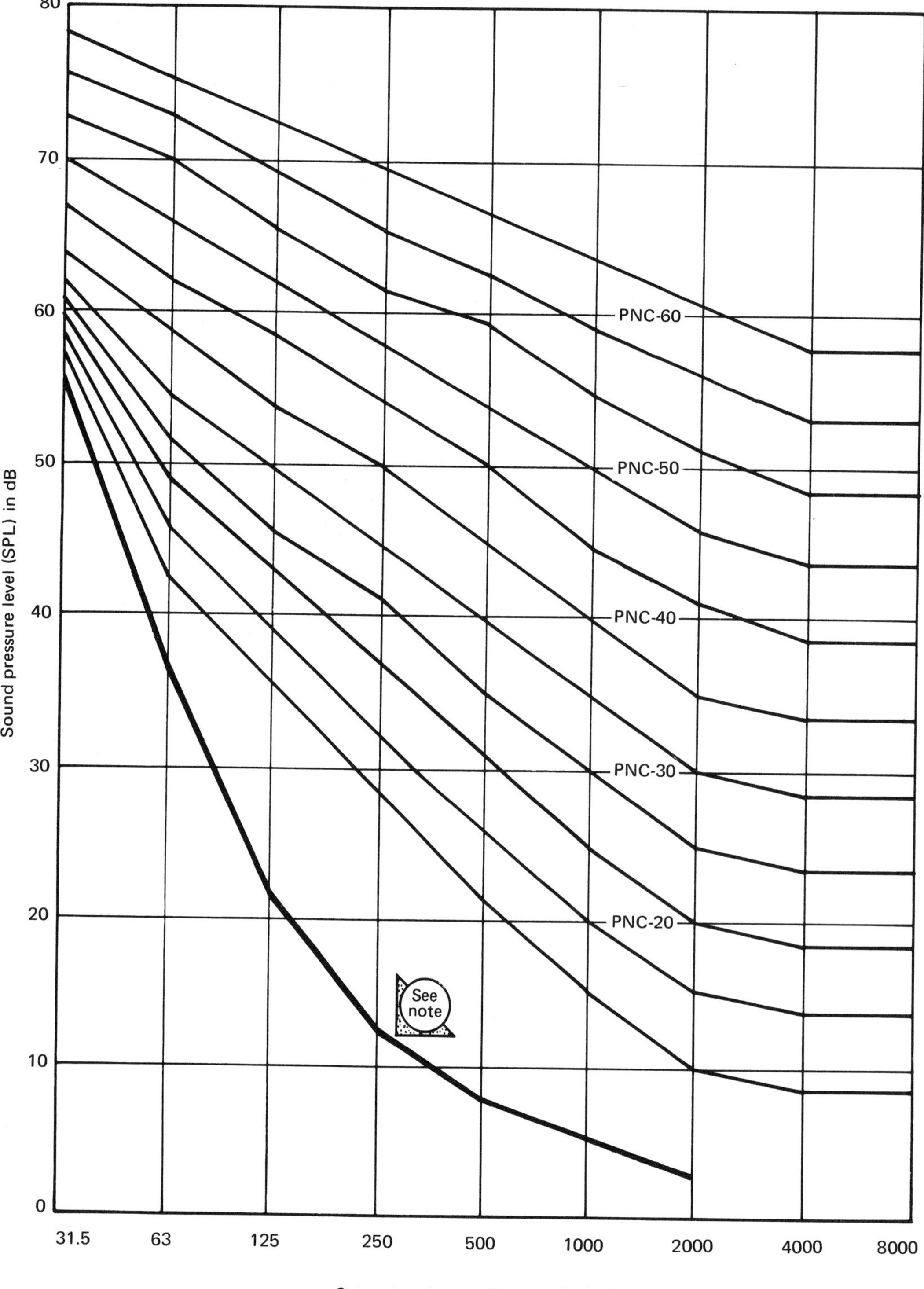

NOTE: Dark curve indicates threshold of hearing for continuous noise (Ref: ***Acoustica*****, Vol. 4, p. 33, Fig. 14, 1964).**

PREFERRED NOISE CRITERIA SOUND PRESSURE LEVEL TABLE*

PNC Curve	Sound Pressure Level, dB								
	31.5 Hz	63 Hz	125 Hz	250 Hz	500 Hz	1000 Hz	2000 Hz	4000 Hz	8000 Hz
PNC-65	79	76	73	70	67	64	61	58	58
PNC-60	76	73	69	66	63	59	56	53	53
PNC-55	73	70	66	62	59	55	51	48	48
PNC-50	70	66	62	58	54	50	46	43	43
PNC-45	67	63	58	54	50	45	41	38	38
PNC-40	64	59	54	50	45	40	35	33	33
PNC-35	62	55	50	45	40	35	30	28	28
PNC-30	61	52	46	41	35	30	25	23	23
PNC-25	60	49	43	37	31	25	20	18	18
PNC-20	59	46	39	32	26	20	15	13	13
PNC-15	58	43	35	28	21	15	10	8	8

*For convenience in using preferred noise criteria data, the table lists the sound pressure levels (SPL's) in decibels for the PNC curves from the preceding page.

Name Index

Subject Index*

*Checklists, example problems, and tables also are listed for easy reference.